Environmental Infrastructure in African History

Examining the Myth of Natural Resource Management in Namibia

Environmental Infrastructure in African History offers a new approach for analyzing and narrating environmental change. Environmental change is conventionally understood as occurring in a linear fashion, moving from a state of more Nature to a state of less Nature and more Culture. In this model, non-Western and premodern societies live off natural resources, whereas more modern societies rely on artifact, or Nature that is transformed and domesticated through science and technology into Culture. In contrast, Emmanuel Kreike argues that both non-Western and premodern societies inhabit a dynamic middle ground between Nature and Culture. He asserts that humans – in collaboration with plants, animals, and other animate and inanimate forces – create environmental infrastructure that constantly is remade and reimagined in the face of ongoing and plural processes of change.

Emmanuel Kreike is Professor of History at Princeton University. He is the author of *Deforestation and Reforestation in Namibia: The Global Consequences of Local Contradictions* (2010) and *Re-Creating Eden: Land Use, Environment, and Society in Southern Angola and Northern Namibia* (2004). Professor Kreike has contributed chapters to several volumes, including *The Nature of Cities* (edited by Andrew C. Isenberg, 2006) and *Social History and African Environments* (edited by William Beinart and Joann McGregor, 2003). He serves on the Executive Committee of the Princeton Environmental Institute (PEI) and the Board of Princeton-in-Africa (PiAF).

Studies in Environment and History

Editors

Donald Worster, *University of Kansas*
J. R. McNeill, *Georgetown University*

Editor Emeritus

Alfred W. Crosby, *University of Texas at Austin*

Other Books in the Series

Donald Worster *Nature's Economy: A History of Ecological Ideas*, second edition
Kenneth F. Kiple *The Caribbean Slave: A Biological History*
Alfred W. Crosby *Ecological Imperialism: The Biological Expansion of Europe, 900–1900*, second edition
Arthur F. McEvoy *The Fisherman's Problem: Ecology and Law in the California Fisheries, 1850–1980*
Robert Harms *Games Against Nature: An Eco-Cultural History of the Nunu of Equatorial Africa*
Warren Dean *Brazil and the Struggle for Rubber: A Study in Environmental History*
Samuel P. Hays *Beauty, Health, and Permanence: Environmental Politics in the United States, 1955–1985*
Donald Worster *The Ends of the Earth: Perspectives on Modern Environmental History*
Michael Williams *Americans and Their Forests: A Historical Geography*
Timothy Silver *A New Face on the Countryside: Indians, Colonists, and Slaves in the South Atlantic Forests, 1500–1800*
Theodore Steinberg *Nature Incorporated: Industrialization and the Waters of New England*
J. R. McNeill *The Mountains of the Mediterranean World: An Environmental History*
Elinor G. K. Melville *A Plague of Sheep: Environmental Consequences of the Conquest of Mexico*
Richard H. Grove *Green Imperialism: Colonial Expansion, Tropical Island Edens and the Origins of Environmentalism, 1600–1860*
Mark Elvin and Tsui'jung Liu *Sediments of Time: Environment and Society in Chinese History*
Robert B. Marks *Tigers, Rice, Silk, and Silt: Environment and Economy in Late Imperial South China*
Thomas Dunlap *Nature and the English Diaspora*
Andrew Isenberg *The Destruction of the Bison: An Environmental History*
Edmund Russell *War and Nature: Fighting Humans and Insects with Chemicals from World War I to Silent Spring*
Judith Shapiro *Mao's War against Nature: Politics and the Environment in Revolutionary China*
Adam Rome *The Bulldozer in the Countryside: Suburban Sprawl and the Rise of American Environmentalism*
Nancy J. Jacobs *Environment, Power, and Injustice: A South African History*
Matthew D. Evenden *Fish versus Power: An Environmental History of the Fraser River*
Myrna I. Santiago *The Ecology of Oil: Environment, Labor, and the Mexican Revolution, 1900–1938*
Frank Uekoetter *The Green and the Brown: A History of Conservation in Nazi Germany*
James L. A. Webb, Jr. *Humanity's Burden: A Global History of Malaria*
Richard W. Judd *The Untilled Garden: Natural History and the Spirit of Conservation in America, 1740–1840*
Edmund Russell *Evolutionary History: Uniting History and Biology to Understand Life on Earth*
Alan Mikhail *Nature and Empire in Ottoman Egypt: An Environmental History*
Gregory T. Cushman *Guano and the Opening of the Pacific World: A Global Ecological History*

Environmental Infrastructure in African History

Examining the Myth of Natural Resource Management in Namibia

EMMANUEL KREIKE
Princeton University

CAMBRIDGE
UNIVERSITY PRESS

CAMBRIDGE
UNIVERSITY PRESS

University Printing House, Cambridge CB2 8BS, United Kingdom

One Liberty Plaza, 20th Floor, New York, NY 10006, USA

477 Williamstown Road, Port Melbourne, VIC 3207, Australia

314-321, 3rd Floor, Plot 3, Splendor Forum, Jasola District Centre, New Delhi - 110025, India

79 Anson Road, #06-04/06, Singapore 079906

Cambridge University Press is part of the University of Cambridge.

It furthers the University's mission by disseminating knowledge in the pursuit of education, learning and research at the highest international levels of excellence.

www.cambridge.org
Information on this title: www.cambridge.org/9781107001510

First published 2013

A catalogue record for this publication is available from the British Library

Library of Congress Cataloging in Publication data
Kreike, Emmanuel, 1959– author.
Environmental infrastructure in African history : examining the myth of natural resource management in Namibia / Emmanuel Kreike, Princeton University.
pages cm. – (Studies in environment and history)
Includes index.
ISBN 978-1-107-00151-0 (hardback)
1. Namibia – Environmental conditions. 2. Human ecology – Namibia – History. 3. Natural resources – Namibia – Management – History. I. Title.
GE160.N3K74 2013
333.7096881–dc23 2012031962

ISBN 978-1-107-00151-0 Hardback

Voor Manus Kreike-Martin

Contents

List of Illustrations

Acknowledgments

This book draws on the knowledge, inspiration, and support of many. Julius Abraham, Kaulipondwa Tuyenikalao Augustaf, Abisai Dula, Philippus Haidima, Kulaumoni Haifeke, Helaliah Hailonda, Nahango Hailonga, Mwulifundja Linekela Haiyaka, Alpheus Hamundja, Hendrik Hamunime, Helemiah Hamutenya, Kanime Hamyela, Shangeshapwako Rachela Hauladi, Juliah Hauwuulu, Israel Hendjala, Francisca Herman, Monika Hidengwa, Petrus Shanika Hipetwa, Matias Kafita, Moses Kakoto, Malita Kalomo, Joseph Kambangula, Gabriel Kautwima, Mathias Malaula, Magdalena Malonde, Islael Mbuba, Petrus Mbubi, Kaulikalelwa Oshitina Muhonghwo, Moses Mundjele, Helivi Mungandjela, Joshua Mutilifa, Kalolina Naholo, Helena Nailonga, Timotheus Nakale, Ester Nande, Paulus Nandenga, Matteus Nangobe, Emilia Nusiku Nangolo, Elisabeth Ndemutela, Werner Nghionanye, Joseph Nghudika, Louisa Palanga, Pauline, Lea Paulus, Marcus Paulus, Vittoria Petrus, Lydia Polopolo, Twemuna Shifidi, Erastus Shilongo, Johannes Shipunda, Joseph Shuya, Selma Tobias, Salome Tushimbeni, Paulus Wanakashimba, and Maria Weyulu generously invited me into their north-central Namibian homes and shared their histories with me. Jackson Hamatwi, a former St. Mary's high school teacher and SWAPO activist, was my cultural, linguistic, and geographical guide, taking me along infrequently trodden paths. I also enjoyed the hospitality and gained much from the insights of Dr. Peter and Jane Katjavivi, Sally and Bishop Kauluma, the late Michael Hishikushitja of St. Mary's Mission at Odibo, and Joseph Hailwa, the Director of the Department of Forestry. The

OMITI survey would have been impossible without the enthusiastic support of the professors and students of Ogongo Agricultural College and the University of Namibia (UNAM), in particular Haveeshe Nekongo, Arne Larssen, and Carlos Salinas. The Namibian Directorate of Forestry, the Dutch Embassy in Namibia, and IBIS-Denmark made the OMITI survey logistically possible. Equally important was the assistance rendered by the very knowledgeable staff of the National Archives of Namibia, especially Werner Hillebrecht and the late Brigitte Lau. St. Mary's Center at Odibo was my home for much of 1991 and 1992 and Ogongo Agricultural College for part of 1993.

I am also grateful to the staff of the Historical Archives of Angola in Luanda, the Archives for Overseas History in Lisbon, the Archives of the Missionary Holy Ghost Congregation in Paris, and the United Evangelical Mission in Wuppertal-Barmen. I would also like to acknowledge the institutional and financial support from Princeton University, Wageningen University, Yale University, and the Social Science Research Council (SSRC), New York.

During the writing phase, I was fortunate to receive intellectual guidance and advice from especially Marius Wessel and Freerk Wiersum at the Department of Forestry, School of Environmental Sciences at Wageningen University; Robert Tignor, Peter Brown, and William Jordan at Princeton University; and William Beinart, Peter Boomgaard, Petra van Dam, Steve Feierman, K. E. Giller, P. Hebinck, Susanna Hecht, Andrew Isenberg, Nancy Jacobs, and Robert Papstein. The comments of Donald Worster, John McNeill, and two anonymous readers commissioned by Cambridge University Press were invaluable. I am also grateful to Eric Crahan, Scott Parris, Kristin Purdy, and Abigail Zorbaugh of Cambridge University Press, as well as Anuj Antony of Aptara Inc., for guiding me and the manuscript to and through the production process. Tsering Shawa, head of the Digital Map and Geospatial Information Center at Princeton University, helped me prepare the map.

Sections of Chapter 5 appeared in "De-Globalisation and Deforestation in Colonial Africa: Closed Markets, the Cattle Complex, and Environmental Change in North-Central Namibia, 1890–1990," *Journal of Southern African Studies* 35(1) (2009), pp. 81–98. A much earlier version of Chapter 2 was published as "Hidden Fruits: A Social Ecology of Fruit Trees in Namibia & Angola, 1880s–1990s," in William Beinart and JoAnn McGregor (eds.), *Social History & African Environments* (James Currey, Ohio University Press, and David Philip, 2003), pp. 27–42. I am thankful to these publishers for allowing me to use these materials in the current volume.

Throughout, my wife Carol and our two children, Manus and Nora, have shared in the intellectual and physical journeys that culminated in the book, which I dedicate to Manus.

Abbreviations

A233	Accessions: Chapman Collection
A450	Accessions: Carl Hugo Linsingen Hahn Papers
AGCSSp	Archives Générales de la Congrégation du Saint-Esprit (Central Archives of the Holy Spirit Congregation, Paris)
AGR	South West Africa Administration: Directorate of Agriculture
AHU	Arquivo Histórico Ultramarino (Archive for Overseas History, Lisbon)
ANC	Assistant Native Affairs Commissioner
ANCO	Assistant Native Affairs Commissioner Ovamboland
AVEM	Archiv der Vereinigten Evangelischen Mission (Archives of the United Evangelical Mission, Wuppertal-Barmen)
BAC	Bantu Affairs Commissioner
BOS	Bantu Affairs Commissioner Oshikango
CEM	Church of England Mission (Anglican)
CNC	Chief Native Commissioner
CNDIH	Centro Nacional de Documentação e Investigação Histórica (National Center for Historical Documentation and Research, Luanda)
CU	Cattle Unit
FMS	Finnish Mission Society
GRN	Government Representative Namakunde (Neutral Zone)
GVO	Government Veterinary Officer
NAN	National Archives of Namibia (Windhoek)
NAO	Native Affairs Ovamboland
NC	Native Commissioner

NCO	Native Commissioner Ovamboland
O/C	Officer Commanding
OMITI	Ovambo Multi-Purpose Investigation for Tree-Use Improvement
OVA	Ovamboland Administration
OVE	Ovamboland Economic Affairs Department
OVJ	Ovamboland Justice Department
RCO	Resident Commissioner Ovamboland
RMG	Rheinische Missionsgesellschaft (Rhenish Mission Society, Wuppertal-Barmen)
SAP	South African Police
Sec.	Secretary (for South West Africa, etc.)
SWA	South West Africa (Namibia)
SWAA	South West Africa Administrator
SWANLA	South West African Native Labour Association
SWAPO	South West Africa People's Organisation
UA	Union (of South Africa) Administration
UGR	Union Government Representative (Neutral Zone)
UNG	Union Government Representative Namakunde (Neutral Zone)
WAT	South West Africa Administration: Directorate of Water Affairs

1

The Ends of Nature and Culture

This book offers a new approach to understanding environmental change that builds on and pushes beyond the idea of a middle ground or middle zone between Nature and Culture.[1] The approach qualifies humans as "architects of Nature" rather than as Nature's rulers or victims, or as its destroyers or champions.[2] Instead of separating the realms of Nature and Culture, the natural and the artifact, or natural resources and technology, this book bridges the dichotomy: the products of both Nature's and Culture's creativity constitute environmental infrastructure. Moreover, whereas current models and approaches are primarily concerned with

[1] White introduced the concept of the "middle ground" to critique the concept of the frontier as a site of incessant conflict between European settlers/Western Culture on the one hand and Amerindians/non-Western Nature on the other hand. See R. White, *The Middle Ground: Indians, Empires, and Republics in the Great Lakes Region, 1650–1815* (Cambridge: Cambridge University Press, 1991). Dove, Sajise, and Doolittle use "middle zone" to denote what lays between the least and the most anthropogenic landscapes; see M. R. Dove, P. E. Sajise, and A. A. Doolittle, "Introduction: The Problem of Conserving Nature in Cultural Landscapes," in M. R. Dove, P. E. Sajise, and A. A. Doolittle (eds.), *Conserving Nature in Culture: Case Studies from Southeast Asia* (New Haven, CT: Yale University Southeast Asia Studies, 2005), p. 8. Agrawal and Sivaramakrishnan refer to agrarian environments as changing hybrid landscapes; see A. Agrawal and K. Sivaramakrishnan (eds.), *Agrarian Environments: Resources, Representation, and Rule in India* (Durham: Duke University Press, 2000), pp. 6–7.

[2] Schama stresses that humans are not, by definition, destroyers of Nature; S. Schama, *Landscape and Memory* (New York: Alfred Knopf, 1995), pp. 9–10. Acknowledging that Amerindians and other "aboriginal" peoples are real agents with a capacity to change their environments for the better or the worse is still controversial; see, e.g., Isenberg, who highlights Amerindian hunting as being partly responsible for the extermination of the bison. See A. C. Isenberg, *The Destruction of the Bison* (Cambridge: Cambridge University Press, 2000).

the outcome of environmental change (i.e., environmental degradation, amelioration, or sustainability) and the origins of its causes (i.e., either natural or anthropogenic), the focus on environmental *infrastructuring* or *environing* advances understanding of the *processes* of environmental change, highlighting the pluralistic and differentiated character of the agency, motivations, and mechanics involved. To be sure, environing should not be construed to be, per definition, an environmentally benign, degrading, or sustainable process.

Although modern societies are thought to construct, manage, and use *artificial* infrastructure (Cultural artifact), indigenous (non-Western) and premodern peoples are held to manage and use *natural* resources. The distinction has largely remained intact, even though the West–non-West and the Nature–Culture dichotomies have been challenged.[3] The idea that Western or "modern" people, armed with science and technology and motivated by a modern market-oriented outlook, actively seeks to harness and replace Nature with human constructs places modernity, the West, and science in the realm of Culture. The complementary idea that non-Westerners or "premodern" people live *by* Nature and *in* Nature assigns premodern, non-Western, and indigenous knowledge to the realm of Nature.[4]

[3] In his path-breaking *Rivers of Empire*, Worster attempts to move beyond both dichotomies. He argues that in the process of subjugating Nature, (human) Culture gets infused by it (in effect creating a hybrid; i.e., the modern U.S. West is deeply shaped by its (semi)arid environment) and applied Wittfogel's hydraulic society model (developed for non-Western and premodern society) to the modern United States; see D. Worster, *Rivers of Empire: Water, Aridity, and the Growth of the American West* (New York: Oxford University Press, 1985), p. 30. For critiques of the Nature–Culture and/or West–non-West dichotomies, see, e.g., White, *The Middle Ground*; W. Cronon, *Nature's Metropolis: Chicago and the Great West* (New York: W. W. Norton, 1991); R. B. Norgaard, *Development Betrayed: The End of Progress and a Coevolutionary Revisioning of the Future* (London: Routledge, 1994); R. White, *The Organic Machine: The Re-Making of the Columbia River* (New York: Hill and Wang, 2000); B. Latour, *Politics of Nature: How to Bring the Sciences into Democracy* (Cambridge, Mass.: Harvard University Press, 2004); Dove, Sajise, and Doolittle, *Conserving Nature in Culture*; A. Agrawal and K. Sivaramakrishnan (eds.), *Agrarian Environments*, pp. 6–10.

[4] On the conflation of the West with the domesticated, civilized, and Culture and the non-West with the uncivilized, wild, and Nature, see H. Ritvo, *The Animal Estate: The English and Other Creatures in the Victorian Age* (Cambridge, Mass.: Harvard University Press, 1987), pp. 16–41. Ritvo argues that this distinction was further elaborated in nineteenth-century English humanitarianism, which was less about protecting animals (or one could add a lesser class of humans, including slaves) against abuse and cruelty, and more about creating and maintaining a social order based on a hierarchical relationship between the civilized classes and subject humans in England as well as across the empire, ibid, pp. 125–288. Two critical edited volumes that highlight the dynamics of historical

The Myth of Culture's Environmental Domination and the End of Nature

The threat of avian and swine flus, HIV-AIDS, malaria, Ebola, and Marburg; the development of penicillin-resistant bacteria alongside the resurgence of tuberculosis; floods, wildfires, and fresh water and fuel shortages; nuclear disasters at Three Mile Island, Chernobyl, and Fukushima; and mounting pollution and global warming – all have cast doubt about the supremacy of Culture's reign over Nature.[5] Until recently, canalized rivers and large dams that generated hydroelectric power and spawned lush irrigated fields, orchards, gardens, and lawns were considered major tools and symbols of the victory of modern industrial society over Nature.[6] The historian Richard White, however, contends that the Columbia River, arguably the most dammed (and thus an artifice and paragon of Culture) and damned (in terms of the destruction of its Nature) river in the most technologically advanced world power, the United States, is in fact neither Culture nor Nature. White calls the Columbia River an

environmental change in Asia reject the conflation of the Nature–Culture and West–non-West dichotomies and its premise of a West–non-West hierarchical causality by focusing on continuities between precolonial and colonial dynamics. See R. H. Grove, V. Damodaran, and S. Sangwan, "Introduction," in R. H. Grove, V. Damodaran, and S. Sangwan (eds.), *Nature and the Orient: The Environmental History of South and Southeast Asia* (Delhi: OUP, 1998), pp. 13–14, and G. Bankoff and P. Boomgaard, "Introduction: Natural Resources and the Shape of Asian History, 1500–2000," in G. Bankoff and P. Boomgaard (eds.), *A History of Natural Resources in Asia: The Wealth of Nature* (New York: Palgrave MacMillan, 2007), p. 4. See also A. Skaria, "Cathecting the Natural," in A. Agrawal and K. Sivaramakrishnan (eds.), *Agrarian Environments*, pp. 265–276.

5 For the 1960s–1970s optimism about the control over infectious diseases and their subsequent (re)emergence, see L. Garrett, *The Coming Plague: Newly Emerging Disease in a World out of Balance* (New York: Penguin, 1994). On floods, see, e.g., A. Kelman, *A River and Its City: The Nature of Landscape in New Orleans* (Berkeley: University of California Press, 2006 [2003]). On water and fuel as scarce resources, see Worster, *Rivers of Empire*; M. Reisner, *Cadillac Desert: The American West and Its Disappearing Water* (New York: Penguin, 1993 [1986]); and J. McNeill, *Something New under the Sun: An Environmental History of the Twentieth-Century World* (New York: W. W. Norton, 2000).

6 See Worster, *Rivers of Empire*; Reisner, *Cadillac Desert*; R. O. Collins, *The Waters of the Nile: Hydropolitics and the Jonglei Canal, 1900–1988* (Princeton: Markus Wiener, 1996); P. P. Howell and J. A. Allan (eds.), *The Nile: Sharing a Resource: An Historical and Technical Review of Water Management and the Economic and Legal Issues* (Cambridge: Cambridge University Press, 1994); R. Coopey and T. Tvedt (eds.), *A History of Water, Volume 2: The Political Economy of Water* (London: I.B. Tauris, 2006); and R. D'Souza, *Drowned and Dammed: Colonial Capitalism and Flood Control in Eastern India* (New Delhi, India: Oxford University Press, 2006).

"organic machine" because, despite the human transformation of the river, humans have failed to unravel its workings as an ecosystem.[7] Although Worster emphasizes the environmental destruction wrought by modern hydraulics, he depicts Nature as far from permanently defeated. Modern dams are doomed because silt accumulates in their reservoirs, reducing their water-holding capacity and eventually breaching the dams. Downstream, the irrigated soils are slowly poisoned by the accumulation of salts.[8] The water consumption of modern irrigated systems is also unsustainable. Moreover, damming, diking, diverting, and canalizing rivers have not domesticated their Nature, as, e.g., the recent history of the Rhine and Mississippi rivers has demonstrated: destructive floods increased as rivers simultaneously were constrained to an ever narrower bed while embankments, dikes, and silt buildup in the channels raised the river beds.[9] Obviously, the Columbia River is not an exception.

The concept of the organic machine can be extended to the entire edifice of modern industrial and postindustrial society. The construction, maintenance, and renewal of its artificial infrastructure is literally fueled and oiled by a single, dangerously finite *natural* resource: hydrocarbons.

7 White, *Organic Machine*.

8 Worster, *Rivers of Empire*.

9 Worster, *Rivers of Empire*. On the Rhine, see M. Cioc, *The Rhine: An Eco-Biography, 1815–2000* (Seattle and London: University of Washington Press, 2002) and D. Blackbourn, *The Conquest of Nature: Water, Landscape, and the Making of Modern Germany* (New York: Norton, 2006). Biggs makes a similar argument for the Mekong Delta, which is almost entirely cultivated but highly vulnerable to floods; see D. Biggs, *Quagmire: Nation-Building and Nature in the Mekong Delta* (Seattle: University of Washington Press, 2010). On the Mississippi, see Kelman, *A River and Its City*. On wetlands in the United States in general, see A. Vileisis, *Discovering the Unknown Landscape: A History of America's Wetlands* (Washington, DC: Island Press, 1999 [1997]). Collins argues that the Nile is not yet fully dominated by humans because Egypt is still short of water. See Collins, *The Waters of the Nile*. On silt, see M. A. Kahlown, A. D. Khan, and M. Azam, "The World's Largest Contiguous Irrigation System: Developments, Successes and Challenges of the Indus Irrigation System in Pakistan," in T. Tvedt and E. Jakobsson (eds.), *A History of Water Volume 1: Water Control and River Biographies* (London: I. B. Tauris, 2006), pp. 35–54. In a study that was originally published in 1955, Carter and Dale hold up the collapse of ancient irrigated civilizations as a warning to modern industrial society. They argue that modern irrigation technology and chemical fertilizer are no alternative to sound soil conservation. See V. G. Carter and T. Dale, *Topsoil and Civilization* (Norman: University of Oklahoma Press, 1974 [1955]). Tvedt and Jakobsson call "[t]he struggle to control water ... a struggle without end" and argue that "[t]he continuing occurrence of vast floods, devastating droughts, and their aftermath, shows that despite 5,000 years of effort, humanity yet has to succeed totally in its attempt to control water." See T. Tvedt and E. Jakobsson (eds.), *A History of Water Volume 1*, pp. ix–xiv.

Thus the artificial infrastructure of modern society, its technology, and its science not only are built literally on the foundations of Nature but also continue to be entirely dependent on Nature for their production and reproduction.[10] Industrial technology therefore did not make human society less dependent on Nature or provide the means to definitively subjugate, destroy, or bring about the death of Nature. Moreover, postindustrial and postmodern society and science's frontier in many ways appear to lay in Nature: biosciences and biotechnology that create Nature–Culture hybrids. White's concept of the organic machine therefore has far-reaching and profound implications because it problematizes both the understanding of modern (agro)industrial and rural and (sub)urban landscapes as Culture, as well as the concept of Culture as an environmental category.[11]

The inability of modern Western science and technology to fully overcome, harness, or destroy Nature was presaged in the colonial era. Following World War II, imperial governments sought to jumpstart development in their colonies through large-scale dam, irrigation, livestock, animal, and human health, and soil, wildlife, and forest conservation projects. The undertakings amounted to a second colonial conquest, but of Nature rather than territory. The success was mixed. The projects resulted in large-scale population displacements; the massive enclosure of resources in parks, forests, and reserves; land alienation; forced labor; forced cultivation; the spread of disease; draconian veterinary measures; taxing soil conservation schemes; and the suppression of fire regimes. The measures elicited resistance and were in many cases frustrated by wars of independence, violence, or a shortage of means.[12] The failure to

[10] Goudsblom not only acknowledges the "industrial age" as the third major ecological transformation brought about by humans but also highlights a critical continuity: the steam and internal combustion engines that mark industrial society are still dependent "on the forces of nature" because its (fossil) fuels are natural. See J. Goudsblom, *Fire and Civilization* (London: Penguin, 1992), pp. 164–174. See also Worster and Reisner, who highlight how modern industrial agriculture depends on water and McNeill, who emphasizes the extent to which modern industrial society is dependent on fossil fuels, see Worster, *Rivers of Empire*; Reisner, *Cadillac Desert*; and McNeil, *Something New under the Sun*.

[11] Zimdahl is pessimistic about the future of modern agriculture. He stresses that modern monoculture makes agriculture ecologically vulnerable, doubts if weeds as a problem can ever be overcome because of herbicide resistance, and identifies major shortcomings of biotechnology. See R. L. Zimdahl, *A History of Weed Science in the United States* (Amsterdam: Elsevier, 2010), pp. 174–201.

[12] For Africa, see, e.g., A. Isaacman, *Cotton is the Mother of Poverty: Peasants, Work and Rural Struggle in Colonial Mozambique, 1938–1961* (Portsmouth, NH:

subjugate non-Western Nature through colonial and postcolonial development projects was attributed to factors emanating from "non-Western" environmental, social, political, cultural, or economic conditions, rather than to any innate weaknesses of the modernization paradigm itself.[13]

The Myth of the Natural

As with Culture, the boundaries and the very category of Nature have equally been questioned. Forests in the Americas, Southeast Asia, and Africa that had been presumed to be primordial have revealed thousands of sites with urban and rural ruins, as well as other signs of human settlement, use, and management. The island forests of Kalimantan are a case in point. From the seventeenth century onward, its Dayak inhabitants engaged in commercial export agriculture and planted and managed fruit trees, creating an anthropogenic forest.[14] The last conserved

1996); R. P. Neumann, *Imposing Wilderness: Struggles over Livelihood and Nature Preservation in Africa* (Berkeley: University of California Press, 2000 [1998]); N. J. Jacobs, *Environment, Power, and Injustice: A South African History* (Cambridge: Cambridge University Press, 2003); and K. B. Showers, *Imperial Gullies: Soil Conservation in Lesotho* (Athens: Ohio University Press, 2005). For North America, see, e.g., Worster, *Rivers of Empire* and White, *The Organic Machine*. For Asia, see, e.g., R. Guha, *The Unquiet Woods: Ecological Change and Peasant Resistance in the Himalaya* (Berkeley: University of California Press, 1989); N. L. Peluso, *Rich Forests, Poor People: Resource Control and Resistance in Java* (Berkeley: University of California Press, 1994 [1992]); R. H. Grove, V. Damodaran, and S. Sangwan (eds.), *Nature and the Orient*; and D'Souza, *Drowned and Dammed*. On fire, see, e.g., S. Pyne, *Vestal Fire: An Environmental History Told through Fire of Europe and of Europe's Encounter with the World* (Seattle: University of Washington Press, 1997) and S. Pyne, *World Fire: The Culture of Fire on Earth* (Seattle: University of Washington Press, 1997). For a discussion of the second colonial conquest as a conquest of non-Western Nature, see E. Kreike, *Deforestation and Reforestation in Namibia: The Global Consequences of Local Contradictions* (Leiden: Brill, 2010), pp. 44–73.

13 Scott discusses the failure of "development" writ large. In *Seeing Like a State* he attributes the failure of imperialist high modernist development projects to "modernity" (i.e., the modern state) itself, irrespective of where it implemented its projects (in the process blurring the West and non-West dichotomy) because it oversimplified reality to make societies and environments "legible" (i.e., controllable and taxable by the state). But in *The Art of Not Being Governed*, Scott shifts his emphasis to the state-evading strategies employed by the subjects/victims of state development as the main reason for the failure of development in the past. See J. C. Scott, *Seeing Like a State: How Certain Schemes to Improve the Human Condition Have Failed* (New Haven: Yale University Press, 1998) and J. C. Scott, *The Art of Not Being Governed: An Anarchist History of Upland Southeast Asia* (New Haven: Yale University Press, 2009).

14 H. Knapen, *Forests of Fortune? The Environmental History of Southeast Borneo, 1600–1880* (Leiden: KITLV Press, 2001), pp. 189–281 and N. L. Peluso, "Fruit Trees and Family Trees in an Anthropogenic Forest: Ethics of Access, Property Zones, and

Edens – including national parks, forest preserves, game sanctuaries, and biological research stations – are typically neither pristine nor natural. Paradoxically, Africa, the ultimate "wilderness" continent, marked by vast rainforests and extensive savannas, and uniquely the last sanctuary of the large wild mammals that once roamed the earth, is also the continent with the longest human history.[15]

Yet, the late twentieth-century Nature–Culture dichotomy casts a long shadow over the conceptualization, analysis, and presentation of environmental change, even as it has been rejected as too static and too linear. The Nature–Culture dichotomy is fundamental to how the dominant paradigms of environmental change – the modernization, declinist, and inclinist paradigms – imagine environmental change. The modernization, declinist, and inclinist paradigms have, in turn, deeply shaped the major approaches used to understand environmental dynamics over time, including the population pressure, political ecology, ecological imperialism, techno-environmental determinism, and globalization/commodification models.[16]

Environmental Change in Indonesia," in P. Squatriti (ed.), *Natures Past: The Environment and Human History* (Ann Arbor: The University of Michigan Press, 2007), pp. 54–102. See also R. L. Wadley, "The Complex Agroforests of the Iban in West Kalimantan and Their Possible Role in Fallow Management and Forest Regeneration;" Deborah Lawrence et al., "Does Tree Diversity Affect Soil Fertility? Findings from Fallow Systems in West Kalimantan;" and W. de Jong, "Forest Management and Classification of Fallows by Bidayuh Farmers in West Kalimantan," in M. Cairns (ed.), *Voices from the Forest: Integrating Indigenous Knowledge into Sustainable Upland Farming* (Washington, DC: Resources for the Future, 2007), pp. 490–501, 502–514, and 515–520, respectively.

15 J. S. Adams and T. O. McShane, *The Myth of Wild Africa: Conservation without Illusion* (Berkeley: University of California Press, 1996 [1992]). See also, e.g., J. Carruthers, *The Kruger Park: A Social and Political History* (Pietermaritzburg: The University of Natal Press, 1995); Neumann, *Imposing Wilderness* (Neumann worked from the premise that African landscapes were anthropogenic); and J. McCann, *Green Land, Brown Land, Black Land: An Environmental History of Africa, 1800–1990* (Portsmouth, NH: Heinemann, 1999), p. 2. Critical for the exposure of the myth of wild Africa is a body of literature that effectively focuses on the physical and spiritual domestication of Africa's challenging environments through a long-term focus on the transformations related to the development of pastoralism, agriculture, and iron technology. See, e.g., J. Iliffe, *Africa: The History of a Continent* (Cambridge: Cambridge University Press, 1995) and D. L. Schoenbrun, *A Green Place, A Good Place: Agrarian Change, Gender, and Social Identity in the Great Lakes to the 15th Century* (Portsmouth, N. H.: Heinemann, 1998). For a more general argument, see C. Merchant, *The Fate of Nature in Western Culture* (New York: Routledge, 2003) and Kreike, *Deforestation and Reforestation in Namibia*, pp. 16–21 and 144–158.

16 For an overview of the discussion, see Kreike, *Deforestation and Reforestation in Namibia*.

The Nature–Culture dichotomy defines human and nonhuman entities and their physical and mental faculties and products as either part of Nature or of Culture and environmental change as occurring along a Nature-to-Culture continuum. Studies of environmental change predominantly set out to prove a hypothesis of environmental degradation (e.g., deforestation – the declinist perspective) or environmental improvement (e.g., reforestation – the inclinist perspective). The very vocabulary used to express environmental change – including such terms and concepts as agricultural revolution, deforestation, domestication and domesticate, anthropogenic, and sustainability – originate in and are derived from the Nature–Culture binary, and its continued use reproduces the dichotomy.[17] Moving beyond the restraints of the Nature–Culture dichotomy therefore requires the development of not only an alternative conceptualization of environmental change but also a new vocabulary.

The persistence of the Nature–Culture dichotomy is manifested, e.g., in the continued practice of making explicit or implicit distinctions between cultural and natural landscapes, between wild and domesticated plants and animals, and between modern Western-style industrial agriculture, horticulture, and forestry and indigenous or traditional natural resources management.[18] But many plants and animals, including feral animals and the so-called semidomesticated plants, do not fit neatly into this binary framework. The same is true for entire past and present environments – e.g., pre-Columbian cultivated landscapes (i.e., the dark earths of the Amazon), lost urban environments (i.e., Palenque), and shifting cultivation and fire-managed landscapes.

17 Dove notes that the use of the term *anthropogenic*, e.g., perpetuates the Nature–Culture dichotomy. See M. R. Dove, "Nature, Society, and Science in Southeast Asia's Grasslands," in M. R. Dove, *Southeast Asian Grasslands: Understanding a Vernacular Landscape: Canonical Readings* (New York: The New York Botanical Garden Press, 2008), p. 4.

18 For example, Denevan in his pathbreaking study fundamentally challenges many of the premises of the Nature–Culture dichotomy and highlights that the environment (i.e., Nature) and technology (i.e., Culture) interact, but, nevertheless considers the two as discrete categories. See W. M. Denevan, *Cultivated Landscapes of Native Amazonia and the Andes* (Oxford: Oxford University Press, 2002 [2001]), p. 129. Similarly, Doolittle, who also contributed a volume to the innovative *Cultivated Landscapes* trilogy, rejects the Nature–Culture dichotomy and the concept of domestication but also slides back at times (e.g., on pp. 24–27 where he defines agriculture very conventionally as involving domesticated plants). See W. E. Doolittle, *Cultivated Landscapes of Native North America* (Oxford: OUP, 2001 [2000]).

In fact, a closer look at many species and landscapes reveals that few, if any, are undisputedly Natural. Shifting cultivation and hunting in one form or another has dramatically shaped the forests and savannas of the Americas, Africa, and Asia. By far the largest concentration of land-based megafauna is located in Africa, but very little of it is found in the "wilderness." To the contrary, Africa's megafauna survives in heavily human-shaped environments: inside human-created and managed national parks, on communal and private farm and ranch lands, and on game ranches. In and beyond the parks, wildlife populations are not only protected from poaching but also often sustained by human-created wells, boreholes, and water reservoirs. The resulting high densities of lions in parks make them hostile environments for competing predators, especially the endangered cheetah. In fact, the largest remaining free-ranging population of the cheetah occurs in Namibia, on commercial ranches and farms in dense bush vegetation, rather than on the expansive savannas of conservation areas.[19] The phenomenon is not unique to contemporary Africa. The nocturnal Southeast Asian tiger, e.g., abandoned its quintessential

[19] On the cheetah, see L. Marker-Kraus et al., *Cheetah Survival on Namibian Farmlands* (Windhoek, Namibia: Cheetah Conservation Fund, 1996). See also R. L. Eaton, *The Cheetah: The Biology, Ecology, and Behavior of an Endangered Species* (New York: Van Nostrand Reinold, 1974), pp. 25–27. This raises the question to what extent the survival of large mammals in Africa is related to animals' adapting to human shaped environments. It is puzzling that the continent with the longest human history is also unique in terms of its large mammal diversity and quantity. If acknowledged, Africa's rich fauna is generally ascribed to human-animal coevolution and explained in terms of a Darwinian arms race between African humans and animals in which humans only gained the upper hand as a result of modern technology (i.e., firearms) in the twentieth century. Human agency, however, may have been more important, not so much through a "natural" conservation ethic and practice but because animals were managed as an environmental resource. Various species of antelopes were held in captivity in ancient Egypt and the same was true for the cheetah throughout history. Were these animals (semi)domesticated? If so, to what extent are current populations consequently "feral"? The surviving cheetah population is very vulnerable to disease because the species has passed through a genetic bottle neck. This is very likely the result of its near extinction in the past, but both domestic and feral populations of many species are also typically marked by genetic bottlenecks as a direct consequence of the selection process of domestication. On the cheetah's genetic bottleneck, see Marker-Kraus, *Cheetah Survival*. On ancient Egypt, see C. A. Spinage, *The Natural History of Antelopes* (New York: Facts of File Publications, 1986), p. 163. On the typical genetic bottleneck among domesticated animals, see J. Clutton-Brock, *A Natural History of Domesticated Mammals* (Cambridge: Cambridge University Press, 1999 [1981]), pp. 30–31. Clutton-Brock argues that the cheetah is, in fact, semidomesticated and would have been a common domesticate had it not been a rare breeder in captivity, see Clutton-Brock, *A Natural History of Domesticated Mammals*, pp. 199–203.

jungle habitat and became a daylight hunter to better stalk its favorite prey of wild pigs and deer, which in turn were attracted by the crops and fruits available in and around the edges of villages and plantations.[20] In the United States, opossum, squirrel, raccoon, deer, and coyote thrive in suburban environments where they are considered pests. Other animals, including the mouse, rat, sparrow, and starling, have adapted to urban environments worldwide, sharing our homes and gardens: they are domestic animals, although they have not been domesticated.[21] The same is true for many insects, including houseflies, mosquitos, fleas, roaches, ants, and bedbugs, as well as numerous microbes.

Post-Domestication Environments

Environments are generally held to constitute Culture when and where humans conquer, dominate, replace, or destroy Nature, gaining control over the environment's production and reproduction and replacing the natural with the artifice. The process of domestication is critical to the transformation from Nature to Culture. A species (or an entire environment) is defined as domesticated once its reproduction and survival are controlled by and dependent on human action.[22] In terms of the domestication of plant and animal species, the focus is overwhelmingly on human control of *sexual* reproduction. With some notable but highly compartmentalized exceptions (root and tuber crops, grafted fruit trees, and

[20] On the tiger, see P. Boomgaard, *Frontiers of Fear: Tigers and People in the Malay World, 1600–1950* (New Haven: Yale University Press, 2001).

[21] J. C. McLoughlin, *The Animals among Us: Wildlife in the Cities: An Appreciation of Inquilines – The Beasts That Share Our Bed and Board* (New York: The Viking Press, 1978).

[22] Clutton-Brock, *A Natural History of Domesticated Mammals*, pp. 30–33 and 130–135. Ingold tones down the domesticate-wild dichotomy and rejects the idea that domesticated species are "made" by humans. Instead, he sees humans as "isolating" and "selecting" breeding populations, see T. Ingold, "Growing Plants and Raising Animals: An Anthropological Perspective on Domestication," in D. R. Harris (ed.), *The Origins and Spread of Agriculture and Pastoralism in Eurasia* (Washington, DC: Smithsonian Institution Press, 1996), pp. 12–24. Goudie sees genetic manipulation as an essential trait of domestication; see A. Goudie, *The Human Impact on the Natural Environment: Past, Present, and Future* (Malden, MA: Blackwell, 2006 [1981]), p. 13. See also J. R. Harlan, J. M. J. de West, and A. B. L. Stemler (eds.), *Origins of African Plant Domestication* (The Hague: Mouton, 1976), pp. 6–10. Higgs emphasizes the fluid boundaries between the categories of wild, domesticate, and feral; pointing out, e.g., that they interbreed within a single species, E. Higgs, "Archaeology and Domestication," in J. R. Harlan, J. M. J. de West, and A. B. L. Stemler (eds.), *Origins of African Plant Domestication*, pp. 29–39.

modern cloning); however, the role of human agency in *asexual* reproduction (in particular vegetative reproduction) has been overlooked. Many important tree species in Africa and elsewhere are propagated through coppice or cuttings.[23] Sexual reproduction facilitates genetic manipulation, which is often considered a critical trait of the domestication process, but vegetative reproduction through cuttings clones the parent genetic material and therefore does not result in genetic improvement. Pollen analysis often cannot distinguish between wild and domesticated varieties and vegetatively reproduced plants are, by definition, underrepresented in pollen samples because they produce little or no pollen.[24] Size is upheld as another indicator of domestication, but relying on size as a

[23] Ironically, for the major Eurasian fruit trees, their vegetative reproduction is what is considered to set them apart as domesticated, see D. Zohary and M. Hopf, *Domestication of Plants in the Old World: The Origin and Spread of Cultivated Plants in West Asia, Europe, and the Nile Valley* (Oxford: Clarendon Press, 1993 [1987]), pp. 134–180. Sauer argues that the technology of planting plants (vegetative reproduction) was acquired by humans before they mastered growing crops through seeds. But it was seed complexes in both the New and the Old Worlds that later spread at the expense of vegetative complexes. In the long run, vegetative reproduction made plants entirely dependent on humans because some plants may lose the capacity to produce seeds and, thus, vegetative complexes were less resilient. See C. O. Sauer, *Seeds, Spades, Hearths, and Herds: The Domestication of Animals and Foodstuffs* (Cambridge: The MIT Press, 1972 [1952]), pp. viii and 24–25, 45–46, 120–121. A 1993 plant breeding textbook heavily privileges sexual breeding and includes only one chapter on asexual breeding, see N. C. Stoskopf with D. T. Tomes and B. R. Christie, *Plant Breeding: Theory and Practice* (Boulder: Westview Press, 1993). A 2002 training manual on vegetative tree reproduction produced by the International Centre for Research on Agroforestry (ICRAF) – an institution that is usually a champion of indigenous knowledge and technology – makes no mention at all of any indigenous practices of reproducing trees through cuttings and only discusses "modern" vegetative reproduction. See H. Jaenicke and J. Beiest (eds.), *Vegetative Tree Propagation in Agroforestry: Training Guidelines and References* (Nairobi: ICRAF, 2002). The contributions to a 1998 ICRAF publication play down indigenous tree planting with one exception and also almost exclusively focus on seeding rather than vegetative reproduction. See J. A. Maghembe, A. J. Simons, F. Kwesiga, and M. Karieya (eds.), *Selecting Indigenous Trees for Domestication in Southern Africa: Priority Setting with Farmers in Malawi, Tanzania, Zambia, and Zimbabwe* (Nairobi: ICRAF, 1998). Bye argues all domesticated plants are cultivated but not all cultivated plants are domesticated: he states that "[i]n many case [sic: cases], cultivated plants have not undergone drastic genetic modification but are propagated, planted and tended." See R. Bye, "The Role of Humans in the Diversification of Plants in Mexico," in T. P. Ramamoorthy et al. (eds.), *Biological Diversity of Mexico: Origins and Distribution* (New York: Oxford University Press, 1993), pp. 707–731 (especially 722).

[24] For pollen and distinguishing wild and domesticate, see M. A. Sowunmi, "Environmental and Human Responses to Climatic Events in West and West Central Africa during the Late Holocene," in F. A. Hassan (ed.), *Droughts, Food and Culture: Ecological Change and Food Security in Africa's Late Prehistory* (New York: Kluwer Academic, 2002), pp. 95–104 (especially 101). Livingstone emphasizes that, with the exception of maize,

determinant is ambiguous at best. Nineteenth-century English cattle breeding was more about size and prestige than about improving livestock: the champion bulls were fat, inbred, unhealthy, and too costly to raise.[25] Moreover, domesticated animals in Bronze Age and Iron Age Europe were significantly smaller than older and contemporary wild forms.[26] Crop improvement research for semiarid environments, especially in the context of global warming, emphasizes drought tolerance traits rather than fruit or seed size, involving crossbreeding with wild varieties and genetic engineering.[27] Finally, the emphasis on sexual reproduction and its control is also related to the representation of Nature in general, and non-Western Natures in particular, as female/female bodies, as illustrated by such expressions as "Mother Nature," "virgin" and "pristine" Nature, "domesticating" Nature, and the "rape" of Nature.[28]

the pollen of cereals and wild grasses in Africa are identical, see A. Livingstone, "Interaction of Food Production and Changing Vegetation in Africa," in J. Desmond Clark and S. A. Brandt (eds.), *From Hunters to Farmers: The Causes and Consequences of Food Production in Africa* (Berkeley: University of California Press, 1984), pp. 22–25. Root crops and vegetatively reproduced crops in general are "invisible" in the archaeological record, see T. Shaw, "Archaeological Evidence and Effects of Food Producing in Nigeria," in J. Desmond Clark and S. A. Brandt (eds.), *From Hunters to Farmers: The Causes and Consequences of Food Production in Africa* (Berkeley: University of California Press, 1984), pp. 152–157. Higgs concludes that archaeological plant remains in most cases do not suffice to distinguish wild from cultivated plants, see E. Higgs, "Archaeology and Domestication," in J. R. Harlan, J. M. J. de West, and A. B. L. Stemler (eds.), *Origins of African Plant Domestication*, pp. 29–39 (especially 34).The case of the eighteenth century decrease of oak pollen in Argyll, Scotland, illustrates the limits of pollen analysis. The phenomenon was a function not of the decline of the oak vegetation as such, but a shift to an oak coppice regime that limited flowering and pollen production, see T. C. Smout, *Exploring Environmental History: Selected Essays* (Edinburgh: Edinburgh University Press, 2009), p. 3.

25 H. Ritvo, *The Animal Estate*, pp. 45–81 and 82–121.

26 A. Jockenhoevel, "Agrargeschichte der Bronzezeit und vorrömischen Isenzeit (von ca. 2200 v. Chr. bis Christi Geburt)," in J. Lüning, A. Jockenhoevel, H. Bender, T. Capelle, *Deutsche Agrargeschichte: Vor- und Frühgeschichte* (Stuttgart: Eugen Ulmer, 1997), pp. 141–261. The same was true in Africa; see A. Gautier, "The Evidence for the Earliest Livestock in North Africa: Or Adventures with Large Bovids, Ovicaprids, Dogs and Pigs," in F. Hassan (ed.), *Droughts, Food and Culture*, pp. 195–207. Gautier also questions if domestic animals have evolved sufficiently to warrant their classification as a separate species; see p. 196 in his text.

27 C. L. L. Gowda et al., "Opportunities for Improving Crop Water Productivity through Genetic Enhancement of Dryland Cops," in S. P. Wani, J. Rockström, and T. Oweis (eds.), *Rainfed Agriculture: Unlocking the Potential* (Wallingford, UK: CAB International, 2009), pp. 133–163.

28 This also highlights the passivity of Nature, just as, e.g., women were seen as sexually passive, whereas agency is conferred on (Western) men and Culture. Science and technology – the tools of Culture – were considered "male" domains. See, e.g.,

Many of the categories and categorizations in the gray zone between Nature and Culture are explicitly or implicitly defined through the domestication principle. When (sexual) reproduction is not human-controlled but the offspring are "merely" protected by humans, the result is referred to as a semidomesticated, cultivated, or anthropogenic resource or landscape. Bairon and Boutrais claim that in the Lake Chad basin of Africa, "domestic" animals have been modified by humans but are not fully dominated; equally, wild animals are not entirely the product of a natural environment, if only because they play a major role in human imagination. Bairon and Boutrais argue that animals constitute an intermediary category between human and natural environments: donkeys and dogs are semi-wild and bees, ants, and termites are semidomesticated.[29] The house cat might fit in the same category.[30] By the same token, colonial officials often considered African villagers' goats, donkeys, dogs,

C. Merchant, *Ecological Revolutions: Nature, Gender, and Science in New England* (Chapel Hill and London: The University of North Carolina Press, 1989); A. Agrawal and K. Sivaramakrishnan (eds.), *Agrarian Environment*, p. 9; and V. S. Jenkins, *The Lawn: A History of an American Obsession* (Washington, DC: Smithsonian Institution Press, 1994), pp. 132–134. On pollen analysis, see R. D. Hill, "Toward a Model of the History of 'Traditional' Agriculture in Southeast Asia," in P. Boomgaard and D. Henley (eds.), *Smallholders and Stockbreeders: Histories of Foodcrop and Livestock Farming in Southeast Asia* (Leiden: KITLV Press, 2004), pp. 19–45. Okafor researched the strong vegetative reproduction potential of many "semi-wild" on-farm fruit trees in Nigeria, but nevertheless recommended seed-based reproduction for large-scale production, see J. C. Okafor, "Case Study: Role of Conservation and Domestication of Minor Woody Forest Plants for Local Communities in Nigeria," in F. T. Last (ed.), *Tree Crop Ecosystems* (Amsterdam: Elsevier, 2001), pp. 7–17.

29 See C. Baroin and J. Boutrais, "Introduction: L'Homme dans son Environnement: Troisième Volet d'un Tryptique," pp. 1–25; P. Roulon-Doko, "Les Animaux dans les Contes Gbaya, republique Centralafricaine," pp. 183–192; C. Baroin, "L'Âne, ce Mal Aimé," pp. 277–298; I. de Garine, "Contribution à L'Ethnologie du Chien dans le Nord du Cameroen et le Sud-Uest du Tchad: Masa, Muzey, Tupri, Kera" pp. 321–348; and I. G. Wichatitsky, "Chasser dans une Société Agraire: À Propos des Duupa du Massif de Poli," pp. 501–523, in C. Baroin and J. Boutrais (eds.), *L'Homme et L'Animal dans le Bassin du Lac Tchad* (Paris, France: IRD, 1999). Jacobsen thinks bees' status as domesticated is ambiguous. He also notes that modern agriculture is largely dependent on pollination by bees; see R. Jacobsen, *Fruitless Fall: The Collapse of the Honey Bee and the Coming Agricultural Crisis* (New York: Bloomsbury, 2008), pp. 9–10 and 53. The "taming" of "wild" animals, including the elephant, was common in Africa but does not fit within the domestication framework; see, e.g., F. Marshall, "The Origins and Spread of Domestic Animals in East Africa," in R. M. Blench and R. C. MacDonald (eds.), *The Origins and Development of African Livestock: Archaeology, Genetics, Linguistics and Ethnography* (London: University College London Press, 2000), pp. 191–221.

30 Ritvo emphasizes that the cat was long vilified because it was only controlled by humans to a very limited degree: it was difficult to breed, often ran away, and hunted its own food. See Ritvo, *The Animal Estate*, pp. 21–23, 116. Clutton-Brock doubts if the cat

and cattle as pests that caused environmental degradation and spread such diseases as scabies, rabies, rinderpest, and lungsickness. In effect, they viewed the animals as semidomesticated or feral.[31] Conceptually, however, semi-wild, semidomesticated, cultivated, and anthropogenic resources and landscapes are often still implicitly regarded as *natural* resources (and part of Nature) and as fundamentally distinct from the Cultural artifice of the land- and cityscapes of industrial and postindustrial societies, thus (unintentionally) (re)embracing the Nature–Culture dichotomy.[32]

A second gray-zone category consists of once domesticated species, resources, and landscapes that escaped from human control over its reproduction and behavior: feral animals and plants (weeds) and cultural landscapes that are "re-conquered by Nature." The pigeons nesting on urban monuments across the globe are feral. The Columbia River's wild salmon is raised in hatcheries. North America's wild turkey populations are overwhelmingly the descendants of domesticated ancestors and are more aptly described as feral. The same is true for the icon of the American "wilderness": the bison. Western Europe's equally iconic "old growth" forests are also more feral than wild: they developed from heavily managed and used medieval forests. For example, Spain's *dehesas* and similar park landscapes dominated by oaks and other nut-bearing trees in the Mediterranean basin were managed intensively for browse (pigs love

counts as a real domesticate, see J. Clutton-Brock, *A Natural History of Domesticated Mammals*, pp. 131–135.

31 Ritvo stresses that the dogs of the lower classes in England were seen as the carriers of rabies and as primitive "cave dogs" (i.e., as the antithesis of the domesticated middle class dogs). On the basis of Ritvo's assessment, one could argue that, in fact, dogs of the lower class were regarded as semi-wild or feral. See Ritvo, *The Animal Estate*, pp. 167–202. See also L. van Sittert and S. Swart (eds.), *Canis Africanis: A Dog History of Southern Africa* (Leiden: Brill, 2008). Colonial officials in Ovamboland in the 1940s and 1950s equally regarded indigenous cattle as dangerous carriers of diseases and as representative of the dark forces of Nature; see Kreike, *Deforestation and Reforestation in Namibia*, pp. 139–160. Like cats, donkeys in southern Africa roamed freely and fed themselves and, moreover, were regarded as pests by the colonial authorities; see Jacobs, *Environment, Power, and Injustice*, pp. 124–126, 180–201; and Kreike, *Deforestation and Reforestation*, pp. 92–101.

32 Doolittle rejects the notion of "domesticated" landscape as not very useful and instead uses the idea of "cultivated" landscapes, see Doolittle, *Cultivated Landscapes of Native North America*, p. 7. Hather rejects the usefulness of the domestication notion for the history of agricultural origins; see J. G. Hather, "The Origins of Tropical Vegeculture: Zingiberacaceae, Araceae and Discoreaceae in Southeast Asia," in D. R. Harris (ed.), *The Origins and Spread of Agriculture and Pastoralism in Eurasia*, pp. 538–550.

acorns), charcoal, tannin, and cork.[33] Many of Europe's revered "wild" trees including oak, walnut, chestnut, and hazel were domesticates.[34] Highlighting the extent to which the categories of wild, domesticate, and feral are ambiguous and socially produced is the fact that the same species can be seen to be wild, domesticated, and/or feral, not only over time, but

[33] Many weeds originated as cultivated plants, see J. Lüning, "Anfänge und frühe der Landwirtschaft im Neolithikum (5500–2200 v. Chr.)," in J. Lüning, A. Jockenhoevel, H. Bender, T. Capelle, *Deutsche Agrargeschichte: Vor- und Frühgeschichte* (Stuttgart: Eugen Ulmer, 1997), pp. 15–139. On pigeons, see R. F. Johnston and M. Janiga, *Feral Pigeons* (New York: Oxford University Press, 1995) and McLoughlin, *The Animals among Us*, pp. 75–102. Schorger argues that the virtually extinct wild turkey populations were reconstituted through the release of domestic turkeys, farm-raised turkeys, and wild turkeys raised in captivity. But the very act of breeding turkeys defines the animals as domesticated and subsequently releasing them into the wild makes them feral at best. He also emphasizes that much of the knowledge about the behavior of wild turkeys is derived from data on domesticated turkeys. Thus, the "wild" turkey of the United States neither behaviorally nor epistemologically fits within the Nature–Culture and wild-domesticated dichotomy. See A. W. Schorger, *The Wild Turkey: Its History and Domestication* (Norman: University of Oklahoma Press, 1966), pp. vii–viii, 417–441. On salmon, see White, *The Organic Machine*, pp. 89–110. For the American bison, see Isenberg, *The Destruction of the Bison*, pp. 175–192. The status of reindeer is also ambiguous: the species was extensively bred in Eurasia until the mid-twentieth century and wild and domestic reindeer co-existed and are considered "a single biological species"; see E. E. Syroechkovskii, *Wild Reindeer* (Washington, DC: Smithsonian Institute, 1995), pp. 235–237. On Western Europe's forests, see C. Watkins (ed.), *European Woods and Forests: Studies in Cultural History* (Wallingford, UK, 1998), J. Howe and M. Wolfe (eds.), *Inventing Medieval Landscapes: Senses of Place in Western Europe* (Gainesville: University Press of Florida, 2002), and T. C. Smout, *Exploring Environmental History*, pp. 53–70 and 87–97. On Spain, see R. J. Harrison, "Arboriculture in Southwest Europe: *Dehesas* as Managed Woodlands," in D. R. Harris (ed.), *The Origins and Spread of Agriculture and Pastoralism in Eurasia*, pp. 363–367. Evans hails weeds as an example of resilient Nature but stresses that at the same time weeds are "intimately connected with culture" for "they remain dependent on human activities for their habitat and, in a very real sense, for their physical and symbolic existence." See C. L. Evans, *The War on Weeds in the Prairie West: An Environmental History* (Calgary, Alberta: University of Calgary Press, 2002), p. xiv. Gissel introduces the term "subnature" to refer to weeds invading urbanscapes as well as to refer to "forms of nature deemed primitive . . . filthy . . . fearsome . . . or uncontrollable" that bear the human imprint, including urban pollution, mud, etc.; see D. Gisse, *Subnature: Architecture's Other Environments* (New York: Princeton Architectural Press, 2009). Gutier states that feral domesticates can regain "wild" traits, see A. Gautier, *The Evidence for the Earliest Livestock in North Africa*, pp. 195–207 (especially 198).

[34] D. Zohary and M. Hopf, *Domestication of Plants in the Old World*, pp. 177–180. For the cultivation of oak and chestnut in pre-modern Europe, see J. Radkau, *Nature and Power: A Global History of the Environment* (Cambridge: Cambridge University Press, 2008), pp. 61–62, and P. Squatriti, "Trees, Nuts, and Woods at the End of the First Millennium: A Case from the Amalfi Coast," in S. G. Bruce (ed.), *Ecologies and Economies in Medieval and Early Modern Europe* (Leiden: Brill, 2010), pp. 25–44.

also at the same time in different places, as is the case with such plants as cacao, mango, citrus, avocado, passion fruit, and guava. Pre-Columbian Amerindians cultivated the guava, sapodilla, American oil palm, Brazil nut, and the dye-producing annatto shrub. But the trees reverted to the "wild" as a result of the post-contact demographic collapse.[35] The feral category has fired the human imagination as little else: witness the fascination with lost cities and Cultures, end-of-civilization narratives (the fall or collapse of empires and post-nuclear holocaust "society"), feral animals and humans (wolf children, "going native"), and the fear of human-created creatures that escape and go wild, from Frankenstein monsters to robots and computers running amok, escaped super bugs, genetically altered crops, and cloned animals.[36]

Fire landscapes constitute a third gray zone. Fire not only shapes existing vegetation but is also critical for the reproduction of environments. Regular burning favors the reproduction of certain species over others. For example, some important economic and charismatic tree species – firs, oak (in Europe and the United States), Transvaal teak (Africa), teak (East Asia), and eucalyptus (Australia) – are favored by the regular occurrence of fires. In fact, fire is necessary for cone sprouting in firs.[37] Fire

[35] N. J. H. Smith et al., *Tropical Forests and Their Crops* (Ithaca, NY: Comstock Publishing Associates, 1992), pp. 38–185, 240–242, 364–369 (annatto), 384–394 (Brazil nut). Higgs emphasizes that it is unknown how many of today's wild animals and plants may have been domesticated in the past, see E. Higgs, "Archaeology and Domestication," in J. R. Harlan, J. M. J. de West, and A. B. L. Stemler (eds.), *Origins of African Plant Domestication*, pp. 29–39. Many dye plants domesticated from classical times became redundant with the end of the nineteenth-century introduction of synthetic dyes and their cultivation collapsed, see D. Zohary and M. Hopf, *Domestication of Plants in the Old World*, pp. 191–194. For examples of weeds that were originally cultivated plants, see A. C. Zeven and P. M. Zhukovsky, *Dictionary of Cultivated Plants and Their Centres of Diversity. Excluding Ornamentals, Forest Trees and Lower Plants* (Wageningen: Centre for Agricultural Publishing and Documentation, 1975), pp. 28–40.

[36] On the fascination with "wild" or "feral" humans, see, e.g., J. V. Douthwaite, *The Wild Girl, Natural Man, and the Monster: Dangerous Experiments in the Age of Enlightenment* (Chicago: The University of Chicago Press, 2002), pp. 11–69, and M. Newton, *Savage Girls and Wild Boys: A History of Feral Children* (New York: Picador, 2002).

[37] R. J. Whelan, *The Ecology of Fire* (Cambridge: Cambridge University Press, 1997 [1995]), pp. 57–134; W. Cronon, *Changes in the Land: Indians, Colonists and the Ecology of New England* (New York: Hill and Wang, 1988 [1983]); and Pyne, *Vestal Fire*, *Fire in America*, and *World Fire*. Vale rejects the position taken by Pyne and others that pre-Columbian America was a cultural landscape as a result of Amerindians burning regimes. Instead he proposes that the Americas constituted neither a natural landscape nor a humanized landscape, but rather something intermediate with some areas having been humanized and others not. See T. R. Vale, "The Pre-European Landscape of the

suppression favors less fire-resistant species and explains, e.g., the gradual replacement of oak by maple in the United States after World War II and selected acacia species over non-thorny species across semiarid Africa.[38] In highland Madagascar, fire is used to maintain tapia (*Uapaca bojeri*) woodlands that serve as the main food for wild silkworms.[39] Most plants that depend on seed for reproduction tend to be more vulnerable to fire than species that rely on sprouting; fire destroys young seedlings, thus favoring sprouting species in competition with seeding species.[40] This may explain some of the bias in the definition of domestication in favor of seed reproduction and against vegetative reproduction. Wildfires that favor sprouting species are supposedly confined to wild or incompletely domesticated landscapes; domesticated landscapes in turn are defined and dominated by seeder or grafted species.

In general, regular fires create a more open landscape; the less dense undergrowth and the grasses and herbs attract grazers and browsers. The resulting environment is favorable to domestic livestock and wildlife. Burning reduces plant biomass but facilitates post-burn regeneration, generally increasing productivity by removing dead grasses that, e.g., shade new growth.[41] Open landscapes and regular burning also facilitate clearing for settlement and farming and suppress cover for predators and habitats for insect disease carriers, including the tse tse fly (causing *nagana* or sleeping sickness), malaria mosquitoes, and ticks.[42] Referring to fire landscapes as anthropogenic environments, however, may overemphasize human agency. Landscapes are affected dramatically by fire, but it is often difficult to assess the exact origins of fire and its outcomes.[43] For

United States: Pristine or humanized?" in T. R. Vale (ed.), *Fire, Native Peoples and the Natural Landscape* (Washington, DC: Island Press, 2002), pp. 1–39.

38 Suppressing fire favors fire-sensitive hardwood plants' invasion, a process that takes place in the United States and Australia; see, e.g., Whelan, *The Ecology of Fire*, pp. 251–273.

39 C. A. Kull, *Isle of Fire: The Political Ecology of Landscape Burning in Madagascar* (Chicago: The University of Chicago, 2004), pp. 116–144.

40 Whelan, *The Ecology of Fire*, p. 180.

41 R. J. Scholes and B. H. Walker, *An African Savanna: Synthesis of the Nylsvley Study* (Cambridge: Cambridge University Press, 1993), pp. 117–121.

42 Kjekhus argues maintaining open landscapes was a conscious malaria management practice; see H. Kjekhus, *Ecology Control and Economic Development in East African History: The Case of Tanganyika, 1850–1950* (London: Heinemann, 1977). On the benefits of burning, see also Kull, *Isle of Fire*, p. 82.

43 See T. R. Vale (ed.), *Fire, Native Peoples and the Natural Landscape*, and R. Boyd (ed.), *Indians, Fire and the Land in the Pacific Northwest* (Corvallis: Oregon State University Press, 1999).

example, humans may set fires intentionally to shape or destroy vegetation, or unintentionally (a campfire gone out of control), or fires may be caused by lightning. Even when and where humans intentionally use and manage fire, the fire dynamics are dependent on such a wide range of factors, including humidity, wind, and soil and vegetation composition, that it is a double-edged sword.[44] In addition, despite its domestication in hearths and internal combustion chambers, fire continues to be an ambiguous and difficult-to-control tool because it can reproduce and sustain itself independently.[45] Even when confined to combustion chambers, human control and understanding of fire and its impact are far from complete. The combustion chemicals from (open and closed) human fires are responsible for severe pollution and are the main cause of global warming and structural global environmental change, exposing the frailty of human dominance over Nature. Moreover, the internal combustion engines are fueled by scarce and finite natural resources. In many ways, the internal combustion engine is thus an organic machine. Both open and internal combustion fires exist in the gray zone between Nature and Culture. As Pyne, a leading historian of fire, concludes, "[a]t best all anthropogenic fire is a hybrid."[46]

The study of environmental dynamics, because of its preoccupation with the *direction* and *cost* (in terms of *outcome*) of environmental change (Nature or Culture, and sustainability, degradation, or improvement), generally has privileged a specific, singular, and homogenous outcome (and baseline) at the expense of the process.[47] In contrast, the concept of

44 Whelan points out that the effects of fire are very complex and that there are many gaps in our understanding of fire. See Whelan, *The Ecology of Fire*, pp. 1–8.

45 Sauer and Goudsblom consider the domestication of fire as critical to the evolution of human civilization (i.e., the domination of Culture over Nature). See C. O. Sauer, *Seeds, Spades, Hearths, and Herds*, pp. 10–11, and J. Goudsblom, *Fire and Civilization*. This is echoed by Radkau; see Radkau, *Nature and Power*, pp. 41–45.

46 On fire, see O. S. Stewart, *Forgotten Fires: Native Americans and the Transient Wilderness*. Edited and with a Foreword by H. T. Lewis and M. K. Anderson (Norman: University of Oklahoma Press, 2002); Pyne, *World Fire*; Pyne, *Vestal Fire* (on fire as a hybrid, see p. 6); S. J. Pyne, *Fire in America: A Cultural History of Wildland and Rural Fire* (Seattle: University of Washington Press, 1997 [1982]); T. R. Vale (ed.), *Fire, Native Peoples, and the Natural Landscape*; R. Boyd (ed.), *Indians, Fire and the Land in the Pacific Northwest*; and C. A. Kull, *Isle of Fire*.

47 Agrawal and Sivaramakrishnan note that the hybridity and plasticity of agrarian landscapes emerge more clearly "when comprehended as something processual;" see A. Agrawal and K. Sivaramakrishnan (eds.), *Agrarian Environments*, p. 6. Tiffen et al. also emphasize the importance of focusing on the process of environmental change, see M. Tiffen, M. Mortimer, and F. Gichuki, *More People, Less Erosion: Environmental Recovery in Kenya* (Chichester, UK: John Wiley and Sons, 1994), p. 29.

mosaic environments acknowledges diversity in the outcome of human-nature interaction. For example, high differentiation, with areas of less intensively used land coexisting with pockets of intensive land use in a dispersed settlement pattern, marked pre-Columbian Maya landscapes.[48] The concept also does not a priori privilege human's or Nature's agency over the other.[49] Highlighting human-Nature *interdependence*, perceiving environmental change as a *differentiating* process, and acknowledging *both* humans and Nature as environmental actors (with social, political, economic, and cultural repercussions) permit the identification of more dynamic and open-ended impacts and outcomes.

A focus on the process of environmental change requires that not only the baseline and the outcome but also the agency of environmental change is differentiated. A limitation of the modernization, declinist, and inclinist paradigms is that they focus on collectives as the subjects and objects of environmental change, imbuing the collectives with organism-like properties. In the modernization and declinist paradigms, populations and forests are respectively the subject and the object of environmental change.[50] The inclinist paradigm emphasizes the practices and knowledge of indigenous communities that are analytically organized into tribes, ethnic groups, and/or clans, rather than individuals, because indigenous

[48] S. L. Fedick, "Introduction: New Perspectives on Ancient Maya Agriculture and Resource Use"; N. P. Dunning, "A Reexamination of Regional Variability in the Pre-Hispanic Agricultural Landscape"; and A. Ford, "Critical Resource Use," in S. L. Fedick (ed.), *The Managed Mosaic: Ancient Maya Agriculture and Resource Use* (Salt Lake City: University of Utah Press, 1996), pp. 2, 53–91, and 297–299, respectively. It is also used by J. Fairhead and M. Leach in their *Misreading the African Landscape: Society and Ecology in a Forest-Savanna Mosaic* (Cambridge: Cambridge University Press, 1996).

[49] G. A. Bradshaw and P. A. Marquet (eds.), *How Landscapes Change: Human Disturbance and Ecosystem Fragmentation in the Americas* (Berlin: Springer Verlag, 2003), especially the introduction and chapters 1, 2, 5, and 18; T. J. Bassett and D. E. Crummey (eds.), *Land in African Agrarian Systems* (Madison: University of Wisconsin Press, 1993), pp. 43–45; J. G. Evans, *Environmental Archaeology: Principles and Methods* (Phoenix Mill, UK: Sutton Publishing, 1999), p. 53. Merchant proposes the idea of a human-nature partnership instead of focusing on the question of whether people or Nature is dominant; see C. Merchant, *Reinventing Eden: The Fate of Nature in Western Culture* (New York: Routledge, 2003), pp. 6, 217, and 223–226.

[50] Nygren, e.g., notes that deforestation studies with a macrostructural focus depict the peasantry as a monolith instead of taking into account class and gender; see A. Nygren, "Development Discourses and Peasant-Forest Relations: Natural Resource Utilization as Social Process," in M. Doornbos, A. South, and B. White (eds.), *Forests: Nature, People, Power* (Oxford, UK, Malden: Blackwell, 2000), p. 25. For detailed case studies that demonstrate the need to differentiate within communities, see A. Agrawal and K. Sivaramakrishnan (eds.), *Agrarian Environments*.

knowledge is seen as a collective body of knowledge.[51] Similarly, the fields of ecology, environmental studies, agriculture, and forestry analytically highlight ecosystems, plant communities, and taxonomic collectives; the lowest significant level of analysis comprises the species and the subspecies rather than the individual.[52] Moreover, traditional Western science tends to atomize the collectivity but analyze the resulting unit as being representative. A single buffalo, e.g., becomes representative of a herd, a species, a genus, or an order, as opposed to being an individual animal.[53] In short, individuals in human, animal, or plant families and species are not appreciated for their unique qualities; instead, each individual is treated as though it constitutes a core sample. Such thinking facilitates extrapolating the results of, e.g., small trial plots to measure soil erosion to larger areas, regions, or even continents. This practice is not only problematic methodologically but also conceptually

51 On the focus on collectives and collective bodies of knowledge, see, e.g., K. F. Wiersum, *Social Forestry: Changing Perspectives in Forest Science or Practice?* (Wageningen: Wageningen Agricultural University, 1999), pp. 67, 81, 84, 96, 134–135; J. F. Kessy, *Conservation and Utilization of Natural Resources in the East Usambara Forest Reserve: Conventional Views and Local Perspectives* (Wageningen: Wageningen Agricultural University, 1998), p. 21; P. Harries, *Butterflies and Barbarians: Swiss Missionaries and Systems of Knowledge in South-East Africa* (Oxford: James Currey, 2007), pp. 155–181 and 218. The idea of indigenous knowledge as a collective body, however, is highly problematic because certain types of knowledge were kept by specific subgroups; see, e.g., E. C. Mandala, *Work and Control in a Peasant Economy: A History of the Lower Tchiri Valley in Malawi, 1859–1960* (Madison: University of Wisconsin Press, 1990), p. xx, and M. Wagner, "Environment, Community and History: 'Nature in the Mind' in Nineteenth Century and Early Twentieth Century Buha, Tanzania," in G. Maddox, J. Giblin, and I. N. Kimambo (eds.), *Custodians of the Land: Ecology and Culture in the History of Tanzania* (London: James Currey, 1996), p. 176. "Tribe" and "ethnic group" suggest primordiality and homogeneity. For critiques, see L. Vail (ed.), *The Creation of Tribalism in Southern Africa* (London: Currey, 1989) and White, *The Middle Ground*, p. xiv.

52 The major exception is pets, which are considered individuals, K. Thomas, *Man and the Natural World: Changing Attitudes in England, 1500–1800* (New York: Oxford University Press, 1996 [1983]), pp. 100–142, and Ritvo, *The Animal Estate*, pp. 226–242. Grove stresses the non-Western origins of the environmental and biological sciences. Western ecological science developed in the colonial context, where the concept of tribe was central (and anthropology/ethno science coevolved with eco-science). This may be one reason why ecology has a much more holistic focus than other sciences. See R. H. Grove, *Green Imperialism: Colonial Expansion, Tropical Island Edens and the Origins of Environmentalism 1600-1860* (Cambridge: Cambridge University Press, 1997 [1995]).

53 Grove notes that the holistic outlook of modern environmentalism stresses the importance of the (eco)system over the individual so strongly that it may devalue the importance of a plant, an animal, or a human as an individual, Grove, *Green Imperialism*, pp. xii–xiii.

because it generalizes and universalizes the specific and homogenizes both objects and subjects of environmental change.[54]

Environmental Infrastructure and Environing

Humans (and nonhumans) are "architects of Nature" because they are environmental actors – they neither simply mold and manipulate Nature as though it were an inanimate and static resource, nor do they merely temporarily "shape shift" Nature without changing its essence. To the contrary, humans work *with* Nature because Nature itself is at once a *medium* and an *actor*.[55] As architects of Nature, humans apply their physical and mental attributes, contributing to and causing environmental change (climate change, e.g.) in interaction with other local, regional, and global human and nonhuman actors, factors, and processes.

The concept of "environmental infrastructure" facilitates a dynamic focus on the gray zone between Nature and Culture. The concept typically and narrowly refers to unambiguously *artificial* infrastructure that serves to clean polluted *natural* resources or to divert surplus *natural* resources (especially water).[56] In this book, the concept of environmental

54 M. Stocking, "Measuring Land Degradation," in P. Blaikie and H. Brookfield (eds.), *Land Degradation and Society* (London: Methuen, 1987), pp. 50–54; T. T. Kozlowski, P. J. Kramer, and S. G. Pallardy, *The Physiological Ecology of Woody Plants* (San Diego, CA: Academic Press, 1991), pp. 222, 236; K. A. Longman and J. Jeník, *Tropical Forest and Its Environment* (Burnt Mill, England: Longman, 1987 [1974]), p. 116; I. Scoones (ed.), *Dynamics & Diversity: Soil Fertility and Farming Livelihoods in Africa: Case Studies from Ethiopia, Mali and Zimbabwe* (London: Earthscan, 2001), pp. 8–10, 14, 24–25; and P. H. Lindert, *Shifting Ground: The Changing Agricultural Soils of China and Indonesia* (Cambridge, MA: The MIT Press, 2000), pp. 19–44. On the need to differentiate among and within human communities, see, e.g., C. C. Gibson, M. A. McKean, and E. Ostrom (eds.), *People and Forests: Communities, Institutions, and Governance*. Cambridge, MA: MIT Press, 2000.

55 Ingold argues that humans do not so much "transform the world" as collaborate with other creatures "in the world's transformation of itself" and emphasizes that Nature is not merely "innate clay" to be molded by humans. See T. Ingold, "Growing Plants and Raising Animals: Anthropological Perspective on Domestication," pp. 12–24 (especially 23).

56 See, e.g., www.njeit.org *and* eb.worldbank.org/external/projects/main?pagePK=64283627&piPK=73230&theSitePK=40941&menuPK=228424. The expression "green infrastructure" has been used to denote a new approach to conservation to connect society with Nature, see M. A. Benedict and E. T. McMahon, *Green Infrastructure: Linking Landscapes and Communities* (Washington, DC: Island Press, 2006). Gadgil and Guha use "ecological infrastructure" to critique the extent to which Marxist analysis has ignored the ecological basis of human societies by an exclusive

infrastructure is expanded to embrace environmental resources in general. The term "infrastructure" not only stresses the utilitarian value that humans ascribe to the environment but also allows room for human and nonhuman agency to shape or reshape the environment, from mental abstraction to physical execution. The qualification of "environmental" highlights the idea that human control, use, and agency are neither absolute nor exclusive. Unlike conventional *artificial* infrastructure (i.e., bridges, roads, and schools), which is controlled, designed, and created by humans to support human activity, environmental infrastructure is not confined to the realm of Culture. Environmental infrastructure may include, e.g., "cultivated landscapes" (landscapes created by fire regimes and/or shifting cultivation) as well as abandoned urban and rural landscapes, anthropogenic soils, coppice woodland, and water management systems. In addition, the concept may include "Cultural landscapes" created by hydraulic societies, as well as urban landscapes and other examples of organic machinery. Moreover, whereas "infrastructure" is commonly held to support *human* society or enterprise, environmental infrastructure may serve human and nonhuman individuals, communities, activities, forces, and processes. A rural landscape of farms, fields, pastures, and woodlands, e.g., or an urban landscape may sustain not only the human populations that live in it but also nonhuman inhabitants, including domestic and wild animals (deer, coyotes, rats, doves, mice, dogs, cats, roaches, ticks, fleas, bedbugs), crops, wild plants, and weeds.

Environmental infrastructure is not merely a singular outcome or an objective, however. It does not permanently and absolutely transform Nature, as per the concept of domestication. Instead, environmental infrastructure results from multiple processes of environmental change and, as is any outcome, is fleeting. Without constant replacement, repair, and maintenance, the "artificial" infrastructure of modern industrial and postindustrial society cracks, corrodes, and collapses, and is covered by waste, water, and weeds. In the 1970s and 1980s, lack of investment in urban infrastructure and pollution triggered fear about the collapse of urban society in the inner cities, causing out-migration to the suburbs to escape the "urban jungle." The idea of the urban jungle refers not only to growing insecurity, lawlessness, and deteriorating infrastructure but also evokes the image of gutted buildings and abandoned, weed-covered

focus on economic infrastructure, see M. Gadgil and R. Guha, *This Fissured Land: An Ecological History of India* (Berkeley: University of California Press, 1993 [1992]), pp. 12–13.

lots.[57] Suburbia, however, has been revealed to be far from an impregnable bastion of Culture: fire, water, weeds, wild animals, and insects engulfed lawns, gardens, and homes.[58] As revealed by the archaeological record, infrastructuring is a continuous (and discontinuous) process of production and reproduction. Successive layers of habitation in, e.g., London, Paris, and Rome, demonstrate how urban infrastructure was incessantly rebuilt and repaired. Where the process was interrupted or ceased altogether, the layers of urban infrastructure were rapidly covered by earth and vegetation, and Culture was erased and forgotten. Such an outcome applied not only to urban landscapes but also to rural "artificial" infrastructure, as seen, e.g., in the case of the hidden remnants of extensive and elaborate water harvesting and irrigation networks that sustained pre-Columbian Amerindian societies.[59]

Without a continuous investment of labor, capital, and knowledge, environmental infrastructure deteriorates: waterholes, wells, reservoirs, and canals silt up; protective homes, storage facilities, palisades, walls, and fences collapse; park landscapes of fruit trees, farms, fields, and fallows are overgrown by the bush or buried by the desert sands. Population decline resulting in the loss of labor and other available resources

[57] On critiques of the city as the embodiment of Culture and the antithesis of Nature, see, e.g., Cronon, *Nature's Metropolis*; A. Isenberg (ed.), *The Nature of Cities: Culture, Landscape, and Urban Space* (University of Rochester Press, 2006), and D. Brantz, "The Natural Space of Modernity: A Transatlantic Perspective on (Urban) Environmental History" and Bernd Herrmann, "City and Nature and Nature in the City," in U. Lehmkuhl and H. Wellenreuther (eds.), *Historians and Nature: Comparative Approaches to Environmental History* (Oxford: Berg, 2007), pp. 195–225 and 226–256, respectively.

[58] On suburbia, see Jenkins, *The Lawn*; A. Rome, *The Bulldozer in the Countryside: Suburban Sprawl and the Rise of American Environmentalism* (Cambridge: Cambridge University Press, 2005 [2001]); and T. Steinberg, *American Green: The Obsessive Quest for the Perfect Lawn* (New York: W.W. Norton & Co, 2006).

[59] On Amerindian water harvesting and irrigation infrastructure, see Denevan, *Cultivated Landscapes of Native Amazonia and the Andes*, pp. 135,169, 185–211, 278–90; Doolittle, *Cultivated Landscapes of Native North America*, pp. 309–452; T. M. Whitmore and B. L. Turner II, *Cultivated Landscapes of Middle America on the Eve of Conquest* (Oxford: Oxford University Press, 2001), pp. 165–227. The archaeologist Cohen notes that every archaeological site or layer "is a snapshot of an ongoing process." See M. N. Cohen, "Introduction," in M. N. Cohen and C. M. M. Crane-Kramer (eds.), *Ancient Health: Skeletal Indicators of Agricultural and Economic Intensification* (Gainesville, Florida: University Press of Florida, 2007), pp. 1–9 (especially 9). The colonial prohibition of slash-and-burn cultivation in the Western Ghats of India led to the region rapidly being overrun by the jungle and wild animals and the abandonment of its villages, see M. D. Subash Chandran, "Shifting Cultivation, Sacred Groves and Conflicts in Colonial Forest Policy in the Western Ghats," in Grove, Damodaran, and Sangwan, *Nature and the Orient*, pp. 674–707.

as a result of war, plague, flight, or migration may lead to disinvestment in environing and the erosion and collapse of environmental infrastructure. Focusing on creating and re-creating environmental infrastructure as a process – that is, on *environing* – underscores McNeill's point that *under*population and population decline are as critical variables as *over*population and population growth in terms of understanding historical environmental dynamics.[60]

Next to urban environments, it is perhaps hydraulic landscapes that most forcefully symbolize human attempts to impose control over Nature.[61] Embankments and canalization turn raging rivers into highways of commerce, dams, and irrigation channels transform deserts into gardens, and dikes and ditches turn water into land. Creating irrigation infrastructure is sometimes accomplished through huge projects that require enormous mobilizations of labor, and sometimes through small-scale projects with village, neighborhood, or household labor. Irrigation might rely on the mechanics of flood waters in rivers or river-fed lakes, as in the cases of the Egyptian Nile and the Chinese Lake Tai. Irrigation may also rely on the mechanics of waterwheels or pumps driven by human, animal, wind, steam, or other means. Regardless, the creation of irrigation works has never been a one-time investment that produced a permanent

[60] McNeill stresses population as an important factor in environmental change, but instead of using a mechanical Neo-Malthusian or Boserupian population pressure model, he focuses on labor supply as a product of population size relative to the labor demand required to maintain a specific human-made environment. He points out that terracing and irrigation systems, e.g., are very labor-demanding in terms of their maintenance and, if the population of a given environment falls below a threshold that it can no longer provide the minimum labor required to maintain the land use system, this will cause environmental collapse. McNeill's is a key argument that has not been given the attention it deserves and the environing argument builds on that notion. See J. R. McNeill, *The Mountains of the Mediterranean World: An Environmental History* (Cambridge: Cambridge University Press, 1992), pp. 5–7, 147–220.

[61] Worster, *Rivers of Empire*; Cioc, *The Rhine*; Blackbourn, *The Conquest of Nature*; Howell and Allan (eds.), *The Nile*; Collins, *The Waters of the Nile*; F. Bray, *The Rice Economies: Technology and Development in Asian Societies* (Berkeley: University of California Press, 1994 [1986]); M. Mufakharul Islam, *Irrigation, Agriculture, and the Raj Punjab, 1887–1947* (New Delhi, India: Manohar, 1997); P. Boomgaard (ed.), *A World of Water: Rain, Rivers and Seas in Southeast Asian Histories* (Leiden: KITLV Press, 2007). See also the multivolume *History of Water* co-edited by T. Tvedt: T. Tvedt and E. Jakobsson (eds.), *A History of Water Volume 1: Water Control and River Biographies* (London: I.B. Tauris, 2006); R. Coopey and T. Tvedt (eds.), *A History of Water, Volume 2: The Political Economy of Water* (London: I. B. Tauris, 2006); T. Tvedt and T. Oestigaard (eds.), *A History of Water Volume 3: The World of Water* (London: I.B. Taurus, 2006); and T. Tvedt and T. Oestigaard (eds.), *A History of Water, Series II, Vol. 1: Ideas of Water from Ancient Societies to the Modern World* (London: I.B. Tauris, 2010).

and enduring outcome in which humans subjugated Nature forever.[62] The need for repairs, maintenance, and upgrades was incessant: reservoirs, canals, and ditches need to be kept free from silt accumulations, and bunds, embankments, and dikes need to be kept in repair; failure to do so leads to collapse.[63] Most precolonial irrigated rice cultivation in Southeast Asia involved "simple earthen diversion structures, many of them temporary and renewed each year after flood damage."[64] The floodwaters that sustained the irrigation systems were both life-giving

[62] Bray, one of the major authorities on the history of irrigated rice cultivation in Asia, suggests that the biggest challenge with establishing irrigated systems is their initial construction and that little or no further investments are required: "[o]nce the fields and watercourses had been built, wet-rice landscapes tended to be ecologically quite stable." But this underestimates the vulnerability of irrigated systems to floods, lack of investment, decay, war, and demographic collapse. See F. Bray, "Instructive and Nourishing Landscapes: Natural Resources, People, and the State in Late Imperial China," in G. Bankoff and P. Boomgaard (eds.), *A History of Natural Resources in Asia*, pp. 205–225 (especially 220). See also Bray, *The Rice Economies.* Here Bray again emphasizes that the huge initial investments in wet rice cultivation yield stable long-term returns. But it is clear from her study that maintaining and repairing hydraulic infrastructure are critical to prevent its collapse. On the Nile and the incessant need and investments for repair and maintenance, see A. Mikhail, *Nature and Empire in Ottoman Egypt: An Environmental History* (Cambridge: Cambridge University Press, 2011), e.g., pp. 3, 11–12, 38–81, 84, 128–169, 170–200, 247–252. See also P. P. Howell and J. A. Allan (eds.), *The Nile*, especially the chapter by P. M. Chesworth, "The History of Water Use in the Sudan and Egypt," pp. 65–79. On Lake T'ai (Tai or Taihu), see J. R. Stuermer, "Polder Construction and the Pattern of Land Ownership in the T'Ai-Hu Basin during the Southern Sung Dynasty" (Ph.D. Dissertation, University of Pennsylvania, 1980). In his global overview of ancient hydraulic systems, Scarborough repeatedly emphasizes the heavy investments required to maintain and repair both small and large scale systems, see V. L. Scarborough, *The Flow of Power: Ancient Water Systems and Landscapes* (Santa Fe, NM: School of American Research Press, 2003).

[63] The colonial engineer Willcocks championed the restoration of the ancient Bengal irrigation infrastructure that had largely disappeared because of a lack of maintenance; see W. Willcocks, *Ancient System of Irrigation in Bengal and Its Application to Modern Problems* (Delhi, India: B.R. Publishing, 1984 [1930]). McNeill discusses the decline of hydraulic systems as the result of the disinvestment of labor resulting from population decline. See McNeill, *The Mountains of the Mediterranean World.*

[64] On precolonial Southeast Asia, see Hill, "Toward a Model of the History of 'Traditional' Agriculture in Southeast Asia," and J. W. Christie, "The Agricultural Economies of Early Java and Bali," in P. Boomgaard and D. Henley (eds.), *Smallholders and Stockbreeders*, pp. 19–45 (especially 30) and 47–67, respectively. Hunt emphasizes that irrigation systems have heavy maintenance requirements: "[o]nce built, keeping these systems operating is in part a social problem. People must organize to provide the technical skill, and the resources, to keep these systems going." See R. C. Hunt, "Communal Irrigation: A Comparative Perspective," in P. Boomgaard (ed.), *A World of Water: Rain, Rivers and Seas in Southeast Asian Histories* (Leiden: KITLV Press, 2007), pp. 187–208 (especially 190–191).

and destructive, constantly depositing silt and eating away at dikes and embankments.[65] Moreover, the irrigation infrastructure had to be constantly re-calibrated and re-invented in the face of new challenges arising from both inside the hydraulic systems themselves and those imposed by outside forces (siltification and salinization, e.g.). Additional factors included conjunctural and structural changes in water levels in rivers and lakes, caused by, e.g., weather fluctuations or climate changes, or up-stream diversions of water for new irrigation works.[66] Moreover, irrigating drylands and silt depositing in waterlands created environments favorable to malaria.[67] Irrigation, therefore, was not an outcome but an ongoing process. Irrigation infrastructure was highly vulnerable to collapse or decline as a result of not only intentional wartime destruction but also disinvestments of labor and resources, e.g., as a result of population flight, demographic collapse, or neglect. In India, the colonial and the postcolonial state actively discouraged investment in the maintenance of the numerous water harvesting reservoirs (tanks) that sustained small-scale irrigation and other water needs of villages.[68] Neglect could make environmental infrastructure nearly impossible to restore, if only

[65] D'Souza emphasizes that alluvial plains are marked by constantly changing water/land boundaries and can only be understood as highly dynamic environments, see D'Souza, *Drowned and Dammed*, pp. 26–30. Silt buildup was a major problem in ancient Mesopotamia, causing irrigation canal beds to be raised thirty feet above the level of the fields and making field drainage an enormous challenge. See J. D. Hughes, *Pan's Travail: Environmental Problems of the Ancient Greeks and Romans* (Baltimore, MD: The Johns Hopkins University Press, 1994), p. 35. The drainage problem may have contributed critically to salinization, which Hughes identifies as a major cause of the collapse of ancient Sumeria.

[66] The ancient Harappan civilization of the Lower Indus Valley provides a dramatic example of the effects of silt buildup, which amounts to twelve meters while the water table has risen by nine meters since the mid-third century BC. See Scarborough, *The Flow of Power*, pp. 140–146.

[67] On silt, marshes, and malaria, see McNeill, *The Mountains of the Mediterranean World*, pp. 73–74, 85–87, 176–177, 229, 234, 263–264, 311–313, 338–346, 350, 354–355; and Hughes, *Pan's Travail*, pp. 82–85, 188–189.

[68] On India, see A. Agarwal and S. Narain (eds.), *Dying Wisdom: Rise, Fall and Potential of India's Traditional Water Harvesting Systems* (New Delhi, India: Centre for Science and Environment, 2005 [1997]) and Mufakharul, *Islam, Irrigation, Agriculture, and the Raj Punjab*, who points out that much of the irrigation infrastructure had declined by the time of colonial rule. On precolonial Southeast Asia, see Hill, "Toward a Model of the History of 'Traditional' Agriculture in Southeast Asia," and Christie, "The Agricultural Economies of Early Java and Bali," in P. Boomgaard and D. Henley (eds.), *Smallholders and Stockbreeders*, pp. 19–45 (especially 30) and 47–67, respectively. See also Hunt, "Communal Irrigation: A Comparative Perspective," in P. Boomgaard (ed.), *A World of Water*, pp. 187–208 (especially 190–191).

for the enormous labor investments required: Sri Lankan villagers in the eighteenth century told a Dutch colonial governor who contemplated the repair of a large old tank that it would take 500–1,000 men more than two-and-a-half years to put it back into operation.[69]

Polder landscapes, with their straight lines of canals and drains, are arguably the preindustrial era's consummate rural environments that symbolize human attempts to control Nature. Polderlands seem to be even more a human artifact than irrigated landscapes because irrigated landscapes remained part-time drylands or wetlands, but the reclamation of polders transformed wetlands into drylands. Often authorized, sponsored, or subsidized by the state in Imperial China and preindustrial Europe, nobles and merchants, monasteries, clans, towns, and villages created most polders, not infrequently by pooling their resources for the costly and risky undertaking. This considerable one-time investment did not suffice, however, because once established, the polder landscapes did not endure by themselves. Dikes, canals, ditches, sluices, waterwheels, and windmills had to be repaired and rebuilt over and over again in the face of decay and destruction. Draining the water from the polders was in and of itself an incessant struggle. Lowering the water tables caused polder soils to compact and subside; silt buildup and the increasing confinement of the remaining waters caused river, lake, and canal beds and water levels to rise. Lowering lands and waxing waters, as well as the destruction wrought by floods, storms, and wars, meant that maintaining polderlands was subject to ever new challenges. In addition, any slackening of vigilance in repairing and maintaining the dikes could be disastrous and cause the loss of the polderlands. In both China and the Low Countries, separate and elaborate administrative, taxation, and executive bodies (e.g., water boards and dike managers) developed to institutionalize the process of repairs, maintenance, and improvement that prevented the water from gobbling up the land.[70] That the creation of polders

[69] Scarborough, *The Flow of Power*, pp. 134–140.

[70] On China, see R. B. Marks, *Tigers, Rice, Silk, and Silt: Environment and Economy in Late Imperial South China* (Cambridge: Cambridge University Press, 1998), pp. 67–95; M. Elvin and L. Ts'ui-jung (eds.), *Sediments of Time: Environment and Society in Chinese History* (Cambridge: Cambridge University Press, 1998), in particular S. Yoshinobu's chapter "Environment versus Water Control: The Case of the Southern Hangzhou Bay Area from the Mid-Tang through the Qing," pp. 135–164; P. C. Perdue, *Exhausting the Earth: State and Peasant in Hunan, 1500–1850* (Cambridge: Harvard University Press, 1987); M. A. Mihelich, "Polders and the Politics of Land Reclamation in Southeast China during the Northern Sung Dynasty (960–1126)" (Ph.D. Thesis, Cornell University, 1979); and Stuermer, "Polder Construction and the Pattern of Land

was a continuous process of producing and reproducing environmental infrastructure is symbolized in the Netherlands' coat of arms, which

Ownership in the T'Ai-Hu Basin during the Southern Sung Dynasty." On polders in Medieval China and Japan, see Bray, *The Rice Economies*, pp. 36–42. On the Low Countries, see W. TeBrake, *Medieval Frontier: Culture and Ecology in Rijnland* (College Station: Texas A&M, 1985); G. P. van de Ven (ed.), *Leefbaar Laagland: Geschiedenis van de Waterbeheersing en Landaanwinning in Nederland* (Utrecht: Matrijs, 1996 [1993]); W. Reh, C. Steenbergen, and D. Aten, *Zee van Land: De Droogmakerij als Atlas van de Hollandse Landschapsarchitectuur* (Wormer: Stichting Uitgeverij Noord-Holland, 2005); A. Verhulst, *Landschap en Landbouw in Middeleeuws Vlaanderen* (n.p.: Gemeentekrediet, 1995); and P. J. E. M. van Dam, *Vissen in Veenmeren: De Sluisvisserij op Aal tussen Haarlem en Amsterdam en de Ecologische Transformatie in Rijnland, 1440–1530* (Hilversum: Uitgeverij Verloren, 1998). Gottschalk demonstrates that the creation and maintenance of polderlands was an incessant and costly struggle against the sea; see M. K. E. Gottschalk, *Historische Geografie van Westelijk Zeeuws-Vlaanderen, vol. 2: Van het Begin der 15e Eeuw tot de Inundaties tijdens de Tachtigjarige Oorlog* (Assen, Van Gorcum, 1958). Loss of polderlands was not limited to the premodern era. See A. Rahman, *Beel Dakatia: The Environmental Consequences of a Development Disaster* (Dhaka, Bangladesh: University Press, 1995) who discusses the partial failure of a massive 1960s–1980s USAID land reclamation project in the Bengal Delta of Bangladesh. Land reclamation was not limited to China and the Low Countries, but also occurred elsewhere on seaboards and in the interior along lakes and rivers – e.g., in the coastal areas of western and southern Europe, the European interior, and in Mexico (Mexico City is in a polder). See, respectively, P. Wagret, *Polderlands* (London: Methuen, 1968 [1959]); Blackbourn, *The Conquest of Nature;* B. Pulsen, "Agricultural Technology in Medieval Denmark," in G. Astill and J. Langdon (eds.), *Medieval Farming and Technology: The Impact of Agricultural Change in Northwest Europe* (Leiden: Brill, 1997), pp. 115–145 (especially 120 for Danish polders); and V. S. Candiani, "Draining the Basin of Mexico: Science, Technology and Society, 1608–1808" (Ph.D. Thesis University of California, Berkeley, 2004). On polders of ancient origin in the Goa region of India, see A. Agarwal and S. Narain (eds.), *Dying Wisdom*, pp. 226–229. On more recent Indian polders, see E. J. James, "Management of Wetlands in Relation to River Basins – A Study from South-West India," in T. Tvedt and E. Jakobsson (eds.), *A History of Water Volume 1*, pp. 437–459. Irrigation systems also spawned institutionalized administrative bodies, sometimes of a centralized nature, sometimes with more decentralized structures. Worster, e.g., argues (after Wittfogel) that irrigation in the nineteenth- and twentieth-century American West led to a high level of centralized state control with the irrigation technicians in a key role: "hydraulic society." Mikhail, on the other hand, argues that irrigation in the Nile Valley until the nineteenth century was highly decentralized and localized with local irrigation experts in control. See, respectively, Worster, *Rivers of Empire* and Mikhail, *Nature and Empire in Ottoman Egypt.* For other contributions to this discussion, see P. Boomgaard (ed.), *A World of Water*; A. Agarwal and S. Narain (eds.), *Dying Wisdom*; and R. Coopey and T. Tvedt (eds.), *A History of Water, Volume 2*. On soil subsidence, see P. J. E. M. van Dam, "Sinking Peat Bogs: Environmental Change in Holland, 1350–1550," *Journal of Environmental History*, 6 (2001), pp. 32–45. In Chinese polders, the use of mud as a fertilizer raised the level of rice paddies as much as 1 cm per year, see G. Zi-tong, "On the Genetic Classification of Paddy Soils in China," in Institute of Soil Science, Academica Sinica, *Proceedings of Symposium on Paddy Soil* (Beijing: Science Press, 1981), pp. 129–138 (especially 129). In Dutch polders, the regular re-digging of drainages and the spreading of the dug-out mud on the fields had the same effect. See depicts a lioness struggling to remain afloat in the water with the motto: "*Ik worstel en kom boven*" ("I struggle and emerge"), stressing polderland reclamation as an endless fight.[71] The 1953 flood that killed almost 2,000 people when northwestern storms breached the dikes brought home the idea that even industrial society's polderland infrastructure had not secured permanent dominance over Nature, and led to a massive investment in the Dutch sea defenses, the Delta Works. Global climate change, with projected sea level increases, however, suggests that the struggle against the water is far from over – in the Low Countries and elsewhere.

Environing in North-Central Namibia

North-central Namibia's twentieth-century history serves as a laboratory to discuss the reconceptualization of environmental change. This book is not about the history of the region per se and as such is not comprehensive. Instead, the case of north-central Namibia offers insights that serve to elaborate, demonstrate, and illustrate how and why environing as an analytical and descriptive tool enhances understanding processes of environmental change beyond the legacy of the Nature–Culture dichotomy. North-central Namibia – known as Ovamboland until the end of South African colonial rule in 1990 – at first glance seems to hold little promise to make the case for the idea of environmental infrastructure and environing. Barren during the long dry season, drought years only serve to accentuate its desert-like landscape. Average rainfall is barely sufficient to ensure a harvest of the most drought resistant grain and staple crop millet. During the dry season, water is so scarce that cattle are driven to remote cattleposts up to a hundred miles away from the villages of their owners, depriving households of dairy, manure, and other cattle products for months on end. The region inundates during the rainy season, sometimes flooding the fields and nearly always transforming the region into a malaria-ridden swamp, isolating villages on higher ground. The region lacks any spectacular "traditional" features (irrigation works) or

G. Wierda, "Rietzeis, Kantzeis or Takkenzaag," July 2008 (paper in the author's possession), pp. 1–2. Wierda references F. Sonneveld, "Het Slootkanteffect in het Klei op Veengebied in Zuid-Holland," *Boor en Spade*, 7 (1954), p. 181.

71 Blackbourn identifies a very similar sentiment in the German North Sea coastal area, expressed in the saying "Wer will nicht deichen, der muss weichen" ("who does not construct/maintains the dikes, has to flee"), see Blackbourn, *The Conquest of Nature*, pp. 121–132. Radkau suggests that other hydraulic societies, including Teotihuacan and China were equally aware of their "ecological vulnerability." He cites an old Chinese saying: "[d]ikes are of no value without people to care for them." See Radkau, *Nature and Power*, pp. 92 and 108–109.

institutions ("custodians of the land") that could be readily identified as environmental infrastructure.[72] Neither has "traditional" land use in north-central Namibia been hailed as a model for sustainable use. To the contrary, colonial and postcolonial observers stressed that Ovamboland was overgrazed, in the throes of deforestation and desertification, and ravaged by droughts and floods.

Moreover, the turbulent history of north-central Namibia also did not hold much promise for the possibility of identifying "traditional" environmental infrastructure. In addition to the droughts, diseases, famines, and floods, the region was plagued by war, violence, and massive population displacements and migrations; some of the tragedies were sufficiently destructive in their own right to erase any environmental infrastructure. Slave raiding in northern Namibia and adjacent southern Angola continued long after the Atlantic slave trade ended; and the dislocation that resulted was exacerbated by cattle raids that served to meet the high demand for cattle by European traders, especially as rinderpest decimated livestock herds in the late 1890s. Rulers and elites needed slaves, cattle, and ivory to acquire guns, ammunition, horses, and other imports. Portuguese and South African colonial conquests (Germany never occupied Ovamboland) in the early 1900s brought more violence and population displacement and, at the onset of World War I, triggered a region-wide famine that decimated the population. Joint Portuguese and South African military operations in 1917 finally "pacified" the region. After World War I, the League of Nations assigned Namibia as a mandate to South Africa, which continued to rule the territory as a United Nation's trusteeship after World War II. Heavy and violent Portuguese taxation and labor demands caused an exodus from the Portuguese side of the border into South African-controlled Ovamboland. The refugees from the Portuguese colony of Angola resettled in a largely uninhabited "wilderness" area between the northern and southern precolonial polities, leading to massive deforestation. The refugees proved to be particularly vulnerable to an early-1930s drought that developed into a famine: starvation was prevented only through colonial food aid. The late 1960s again found

[72] On "custodians of the lands" and managed environments in East Africa, see J. Ford, *The Role of Trypanosomiases in African Ecology* (Oxford: Clarendon Press, 1971); Kjekhus, *Ecology Control and Economic Development in East African History*; J. Giblin, *The Politics of Environmental Control in Northeastern Tanzania, 1840–1940* (Philadelphia: University of Pennsylvania Press, 1992); and G. Maddox, J. Giblin, and I. N. Kimambo (eds.), *Custodians of the Land: Ecology and Culture in the History of Tanzania* (London: James Currey, 1997).

the region's population drawn into war, leading to new refugee movements back and forth across the border as Portugal and South Africa violently suppressed home-grown liberation movements in their territories. The independence of Angola in 1975 escalated the violence on both sides of the border. War in Ovamboland subsided only in 1990 when the country gained its independence, but violence continued along the border until the conflict in Angola finally ended in 2002. In addition to violence and drought, colonial policies and economic changes also had a dramatic impact on the region's society and its environment. The enforcement of the colonial border between Angola and Namibia inhibited communications and trade, notably seasonal transhumance cattle movements from Ovamboland to dry season pastures north of the border. Ovamboland and the adjacent Lower Kunene region of Angola also became major sources of migrant labor for the mines and white farms in Namibia and South Africa. The World War II industrialization of South Africa dramatically increased the demand for migrant labor from the region and deeply affected Ovamboland's society, economy, and environment. The increase in migrant labor wages facilitated the wide-spread adoption of cattle- and donkey-powered steel plows, allowing households to cultivate larger fields. Wage remittances became a major source of income for most households. Migrant labor also raised male absenteeism, which changed social relations and land use in the villages as women, children, and the elderly were forced to share more of the burden of crop cultivation, livestock herding, and farm maintenance. Women became the de facto land managers even as men successfully reinforced absentee de jure land ownership. Although the border and veterinary restrictions affected pastoralism and the colonial administration alienated seasonal pastures to create the Etosha Game Reserve on the southern edge of Ovamboland (which severely limited hunting), the region escaped the introduction of many other dramatic conservation policies that were implemented elsewhere in colonial Africa. Large conservation projects involving livestock culling or soil conservation were never introduced in Ovamboland, and many conservation rules and practices implemented by South Africa elsewhere in its "native reserves" were not enforced by the district's officials until the late 1950s. This is important because contestations over colonial conservation practices often bring into focus indigenous environmental use and management as captured in colonial documentation.

Fortunately, there is no dearth of sources for Ovamboland. Quite the contrary, archival sources for north-central Namibia are exceptionally abundant and rich. Three colonial administrations (German, Portuguese, and South African) and five missionary societies (the Church of England Mission, the Finnish Mission Society, the Rhenish Mission Society, the Holy Ghost Fathers, and the Missionary Oblates of Maria Immaculate) produced detailed documents about the region and most of the collections are well preserved and accessible. Moreover, the long-serving Native Commissioner for Ovamboland, C. H. L. Hahn, built a large personal archive that included, among other items, his diaries and photographs that are preserved in the National Archives of Namibia in Windhoek. The thick archival documentation on Ovamboland may be partially a product of Namibia's status as a mandate and later trusteeship, which required the South African administration to submit regular reports to the League of Nations and the United Nations, respectively. Moreover, Ovamboland and the adjacent Kunene region of Angola were by far the major sources of migrant labor to the mines and farms of Namibia because the Germans had decimated the African population of central Namibia.

Field research further added to the available historical data. With independence in 1990, the new democratic government strongly encouraged field research in Ovamboland – which facilitated oral research projects, including the author's. With the invaluable assistance of Jackson Hamatwi, a former Anglican mission high school teacher turned South West African People's Organisation (SWAPO) activist who knew all the back roads and tracks in Ovamboland (he smuggled supplies for guerillas past South African roadblocks and patrols), the author was able to interview one or two old men and one or two old women in each of thirty villages spread throughout the Oukwanyama subdistrict of Ovamboland, some close to the single tarred road, others very remote.

The lack of post-independence baseline data on the agriculture and the environment also stimulated numerous household and agricultural surveys as well as a census. In collaboration with the University of Namibia (UNAM), the faculty, staff, and students of Ogongo Agricultural College, and the Ministry of Agriculture, and with financial and logistical support from Dutch, Danish, and Swedish aid organizations and various international and local non-governmental organizations (NGOs), the author organized the Ovamboland Multi-Purpose Investigation for Tree-Use Improvement (OMITI) survey. After a training workshop, teachers, students, and staff of Ogongo Agricultural College surveyed approximately 400 households in 100 villages selected according to criteria that ensured representation of villages from a range of characteristics, including the district to which they belonged, "ethnic" and religious affiliations, and access and distance to such infrastructure as tar, gravel, or dirt roads, piped water or canals, schools, hospitals, shops, and towns, and the international border. At least one member of each of the two-member survey teams was from the region and spoke the local dialects of Oshivambo. All team members were advanced students in either agriculture or forestry. The questions in the survey covered many themes and topics, from household composition and the number of animals a household possessed, to environmental resource management, use and tenure; and the respondents were invited to comment on the present as well as the past (i.e., childhood memories). The author designed the questions based on his experience conducting oral history research in the region during the previous year and after having researched the various archives for two years. Many questions sought quantitative data to follow up on information acquired through the largely qualitative oral histories.

The reconceptualization of environmental change with humans as architects of Nature rather than as its masters or subjects is inspired by how the inhabitants of north-central Namibia cast environmental change in their narratives: as the making and unmaking of farms, fields, and villages; and *oshilongo*, an environment or landscape that has been enhanced with an environmental infrastructure that sustains concentrations of humans, animals, and plants throughout the year. Obviously, this produces a human-centered bias, but it was and is not a linear narrative of the human conquest of Nature, or a static story of a balance between humans and Nature. Interviewees did not portray *oshilongo* as a static and final outcome: instead, they emphasized that without sustained heavy investment of labor, capital, and ideas, the area quickly transformed into *ofuka*, a landscape that lacks the environmental infrastructure capable of supporting a larger population of humans and animals through the dry season and drought, and flood and famine years. Making *oshilongo*, therefore, was and is a continuous process: *environing*.[73] Placing this conceptualization of the process of environmental change in the context of a variety of regional and global bodies of literature suggests that the framework is not merely an example of localized indigenous knowledge and narratives but provides an analytical and descriptive framework for understanding environmental change that moves beyond the linearity and homogeneity of existing models.

[73] E. Kreike, *Re-creating Eden: Land Use, Environment, and Society in Southern Angola and Northern Namibia* (Portsmouth, NH: Heinemann, 2004).

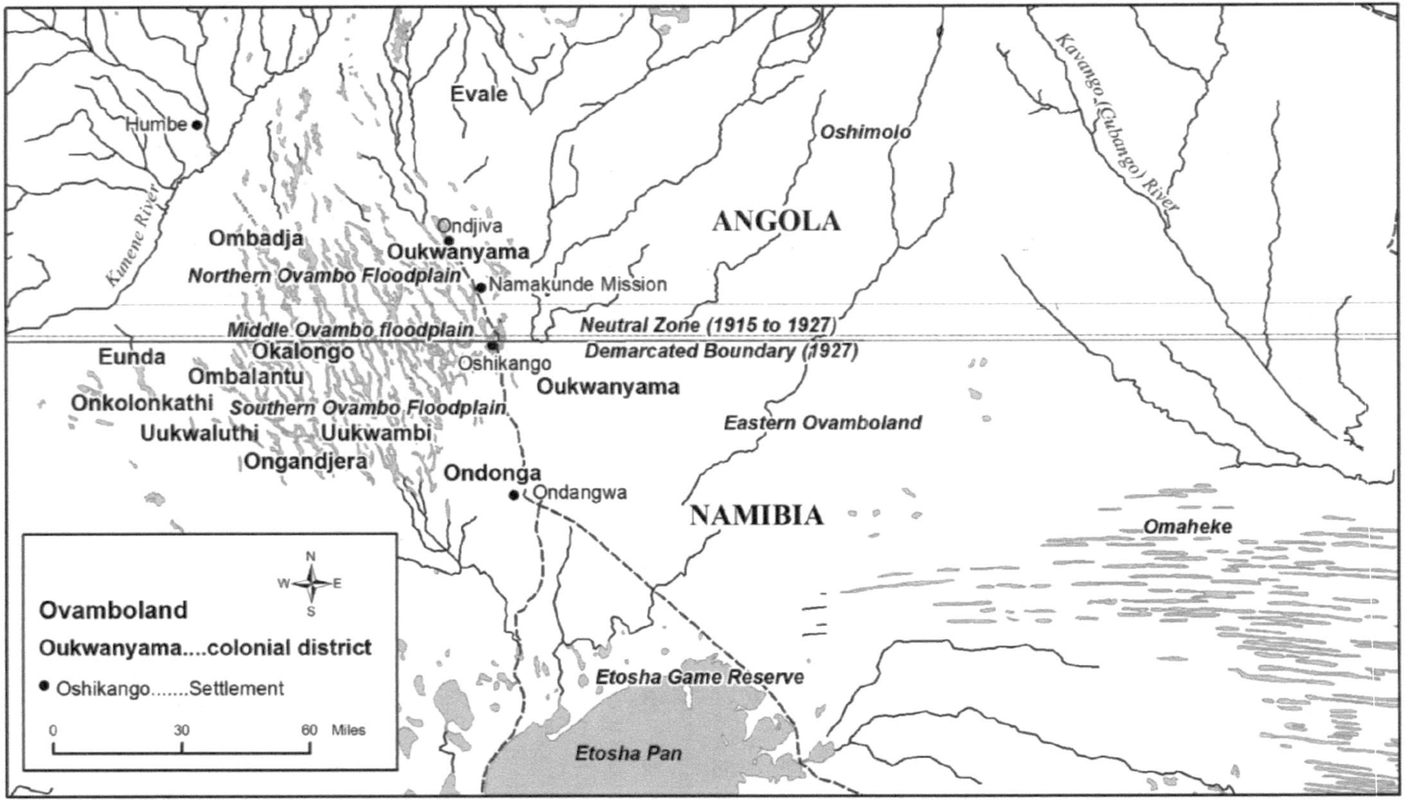

Map of Ovamboland.

Chapter 2 discusses how and why the concept of environmental infrastructure highlights indigenous "natural" resource management and use from a perspective beyond the purview of the Nature–Culture dichotomy. Even as colonial and postcolonial officials and experts considered the enormous fruit trees that dominate north-central Namibia's landscape to be mere wild and wilderness trees, the trees were intimately associated with human settlement. As humans colonized the sparsely inhabited areas along the modern Angolan-Namibian boundary, fruit trees advanced accordingly.

Chapters 3, 4, and 5 discuss the processes of environing at (respectively) the field/farm, village, and landscape levels. "Dark Earths" (Chapter 3) highlights how humans created, maintained, and re-created farms, fields, and even soils. "Water and Woodland Harvesting" (Chapter 4) demonstrates that neither north-central Namibia's water, nor its woodland resources were *natural* resources. The main dry-season source of water for the majority of north-central Namibia's population consisted of human-dug water holes with limited yields. Equally, the intense demand on woodland resources could be sustained only through coppice management. "Browsing and Burning Regimes" (Chapter 5) explores the extent to which Southern Africa's ubiquitous bushlands were neither natural nor simply degraded, but heavily shaped by human-managed browsing and burning.

Chapter 6, "Valuing Environmental Infrastructure," argues that valuing environmental resources – most notably expressed through the introduction of land fees in colonial Ovamboland – was not merely a mechanistic outcome of population pressure that created land scarcity and/or "the market" putting a price on the land. Instead, paying land fees also reflected the perceived increase of a farm plot's value as a result of continuous investments in the creation and maintenance of environmental infrastructure.

"Science and the Failure to Conquer Nature" (Chapter 7) demonstrates that "modern" Western technology is in fact *environmental* rather than *artificial* infrastructure. In north-central Namibia in the early 1970s, a hydraulic system with pipelines and canals that supplied water from the Kunene River throughout the year made urbanization and irrigation possible, but a severe drought in the early 1990s proved that the scientific hydraulic technology had failed to make the region immune to the vagrancies of Nature.

2

Architects of Nature

The analysis and the description of environmental dynamics in Namibia in particular and in the non-Western world in general remain largely constricted by the legacy of the Nature–Culture dichotomy. Notably, even as the role of indigenous environmental agency is increasingly acknowledged, the ensuing processes and outcomes of environmental change and the motivations for change continue to be confined to the realm of Nature by both its distractors and its advocates. For example, even when colonial and postcolonial officials and experts reported that such north-central Namibian fruit trees as the marula occurred almost exclusively on farms, they nevertheless continue to be categorized as wild trees from the wilderness (i.e., relics from natural vegetation).[1] And the southern African marula is far from exceptional in this respect. As a

[1] The marula (*Sclerocarya birrea*) is a common on-farm fruit tree across southern Africa. Carbonized remains of its nut were present at the medieval archaeological site at Mapungubwe in South Africa. See T. Shaw, "Early Crops in Africa: A Review of the Evidence," in J. R. Harlan, J. M. J. de West, and A. B.L. Stemler (eds.), *Origins of African Plant Domestication*, pp. 107–153 (especially 114–115). On the marula, see also K. B. Wilson, "Trees in Fields in Southern Zimbabwe," *Journal of Southern African Studies*, 15 (1989), pp. 369–383. For marula in Malawi, see A. C. Chikuni, "Conservation Status of Mopane Woodlands in Malawi: A Case Study of Mua-Tsanya Forest Reserve," in L. J. G. van der Maesen, X. M. van der Burgt, and J. M. van Medenbach de Rooy (eds.), *The Biodiversity of African Plants* (Dordrecht: Kluwer, 1994), pp. 250–258. Marula and two other major fruit trees common on farms in north-central Namibia, *Berchemia discolor* and *Diospyros mespiliformis* also occur in the southeastern region of Kenya. See P. M. Maundu, "Utilization and Conservation Status of Wild Food Plants in Kenya," in L. J. G. van der Maesen, X. M. van der Burgt, and J. M. van Medenbach de Rooy (eds.), *The Biodiversity of African Plants*, pp. 678–683.

rule, valuable indigenous fruit trees in the tropical regions of Asia, the Americas, and especially Africa are considered to be wild trees. Even when an association with human settlement is acknowledged or if human agency is recognized in propagating the trees, the trees are nevertheless regarded at best as semidomesticated (i.e., advancing tentatively from Nature and the wild to Culture and the domesticated, from part of Nature to part of Culture). At best, they are subject to natural resources management.[2] The African oil palm tree (*Elaeis guineensis*) has been

[2] Ingold notes that trees have been relatively neglected in the literature on domestication because they do not fit well into the orthodox wild-domesticate discussion, see Ingold, "Growing Plants and Raising Animals: Anthropological Perspective on Domestication," pp. 12–24 (especially 22). Leakey and Newton venture that traditionally important woody tropical plants are "virtually undomesticated [and] ignored by the 'green revolution'" and many species they discuss "can . . . at best be described as semi-domesticates." See R. R. B. Leakey and A. C. Newton (eds.), *Tropical Trees: The Potential for Domestication and the Rebuilding of Forest Resources* (London: HMSO, 1994), pp. 3–5. Sinclair et al. emphasize that there are many reports about local tree utilization "[but] it is often unclear whether these refer to the exploitation of wild vegetation or to controlled husbandry" and although there are many reports also about protection and use of trees in dryland Africa "very few documented examples of indigenous tree planting" are available, see F. L. Sinclair, I. Verinumbe, and J. B. Hall, "The Role of Domestication in Agroforestry," in R. R. B. Leakey and A. C. Newton (eds.), *Tropical Trees: The Potential for Domestication and the Rebuilding of Forest Resources*, pp. 124–136 (especially 125). Seme et al. note that many African fruits "grow wild or are partially domesticated." See E. N. Seme, P. W. Chahira, and J. K. Kemel, "*Ex Situ* Conservation of African Crop Germplasm," in A. Putter (ed.), *Safeguarding the Genetic Basis of Africa's Traditional Crops* (n.p.: CTA/IPGRI, 1992), pp. 101–103 (especially 102). Smith et al. point out that it is often very hard to draw a distinction between domesticates and feral species, see Smith et al., *Tropical Forests*, pp. 49–185, 415–417. They also note that the tropical zones of Asia and the Americas are the richest sources of domesticated fruits, but add that "surprisingly Africa and Madagascar contain only a limited diversity of tropical fruit trees." Smith attributes this to humans having transformed the forest vegetation more intensely in Africa, see Smith et al., *Tropical Forests*, pp. 186–187 and 418. But he adds that the relatively small number of domesticated crops in Africa may be caused by the lower species diversity of tropical Africa with the caveat: "[o]n the other hand, further research may well reveal that Africans cultivate or have semi-domesticated a far broader range of plant species than was hitherto suspected" (418). Asia alone counts more than 100 species of tropical fruit that are cultivated and another 100 species of fruit trees are gathered in the forest. The situation is similar in the tropical Americas, less so in the tropical regions of Africa. But it includes the oil palm which spread through migrations across Africa and, because of the slave trade, to the Americas, see Smith et al., *Tropical Forests*, pp. 186–238, 264–315. Smith et al. also point to the potential of little-studied nut producing trees, Smith et al. *Tropical Forests*, pp. 371–413. The tropical forest fruit basket includes a good number of domesticates Smith et al. acknowledge, but many others state "linger on the threshold of domestication" (423). Compare Pursglove, who contests the notion that Africans have been less effective in terms of domestication and stresses that more plants have been domesticated in Africa than is generally accepted,

associated at least in part with human activities in West Africa for almost 4,000 years, but is still considered only semidomesticated. Other African trees that are often closely associated with human settlement but are still considered "wild," "semi-wild," or "semidomesticated" include the ana tree (*Faidherbia albida*), baobab (*Andansonia digitata*), moringa and parkia species (including *Parkia biglobosa* or *Néré*), kola nut (*Cola acuminate* and *Cola nitida*), Shea butter (*Butyrospermum paradoxum*), and

see J. W. Pursglove, "The Origins and Migrations of Crops in Tropical Africa," in J. R. Harlan, J. M. J. de West, and A. B. L. Stemler (eds.), *Origins of African Plant Domestication*, pp. 291–309. Zeven and Zhukovsky identify 80 "cultivated" trees and shrubs in Africa compared to 114 in China/Japan and listed a total of 276 domesticates for Africa. This places Africa as a continent in third place after Indochina/Indonesia with 303 and China/Japan with 284. Africa thus is only thirteen species short of second place and twenty-seven short of first place. See Zeven and Zhukovsky, *Dictionary of Cultivated Plants and Their Centres of Diversity*. In nine out of twenty-four case studies of exploited "forest" trees in Latin America, the trees were associated with human-managed space and most of these were actively managed, protected, transplanted, cultivated or planted, and/or occurred in gardens. See M. N. Alexiades and P. Shanley (eds.), *Productos Forestales, Medios de Subsistencia y Conservación: Estudios de Caso sobre Sistemas de Manejo de Productos Florestales No Maderables: Volumen 3 – America Latina* (Jakarta, Indonesia: CIFOR, 2004). In Peru, 4,400 indigenous plants are used of which 1,700 are cultivated and 182 domesticated. Of the 182 native domesticates, 58 are fruit trees; see A. B. Egg, *Perú: Diez Mil Anos de Domesticacíon* (Lima: UNDP, 2003). Starch-producing palms are often referred to as sago palms although sago is a specific palm that is very common in Asia and Americas and is closely associated with humans. Where they occur outside of swamps, they often have been (vegetatively) propagated by humans and are referred to as "semi-cultivated." See R. Ellen, "The Distribution of *Metroxylon sagu* and the Historical Diffusion of a Complex Traditional Technology," in P. Boomgaard and D. Henley (eds.), *Smallholders and Stockbreeders*, pp. 69–105. A book on forest products in Asia presents twenty-one case studies of what the authors call "managed wild resources": in seven cases they classify the "dominant" form of management as "wild," in two cases as "wild/cultivated," in a single case as "wild/managed," in another two cases as "managed/cultivated," and in six cases as "cultivated." See K. Kusters and B. Belcher (eds.), *Forest Products, Livelihoods and Conservation: Case Studies of Non-Timber Forest Product Systems, Volume 1 Asia* (Jakarta, Indonesia: CIFOR, 2004). Agroforestry practices with trees grown in fields in sequence or interplanted with annual crops have a long history in China. See N. K. Menzies, *Forest and Land Management in Imperial China* (Houndsmills, Basingstoke, Hampshire, UK: MacMillan, 1994), pp. 89–110. Contributors to a 1997 workshop on tree domestication called for a much wider definition of domestication that abandoned the fixation on genetic improvement, but were still hesitant to discuss indigenous tree management in terms of domestication, see J. M. Roshetko and Dale O. Evans (eds.), *Domestication of Agroforestry Trees in Southeast Asia: Proceedings of a Workshop, Yogyakarta, Indonesia, November 4–7, 1997* (Morrilton, AK: Winrock International & ICRAF, 1999). Shipek noted that the Kumeyaay Indians of southern California planted oak groves and various other trees and shrubs; see F. C. Shipek, "An Example of Intensive Plant Husbandry: The Kumeyaay of Southern California," in D. R. Harris and G. C. Hillman (eds.), *Foraging and Farming*, pp. 159–160.

the African pear tree (*Dacryodes edulis*).[3] Kalahari hunter-gatherers use fire to favor and protect, transplant, and even cultivate select "wild" plants, including corky monkey orange (*Strychnos cocculoides*), sour berry (*Grewia* species), and mangetti (*Ricinodendron rautanenii*).[4] North

[3] On African oil palm, see Sowunmi, "Environmental and Human Responses to Climatic Events in West and West Central Africa during the Late Holocene," pp. 95–104; T. Shaw, "Early Crops in Africa: A Review of the Evidence," in J. R. Harlan, J. M. J. de West, and A. B. L. Stemler (eds.), *Origins of African Plant Domestication*, pp. 107–153; and Smith et al., *Tropical Forests*, p. 425. See also M. Posnansky, "Early Agricultural Societies in Ghana," in J. Desmond Clark and S. A. Brandt (eds.), *From Hunters to Farmers: The Causes and Consequences of Food Production in Africa* (Berkeley: University of California Press, 1984), pp. 147–151; Zeven and Zhukovsky, *Dictionary of Cultivated Plants and Their Centres of Diversity*, pp. 108–128 (they list 80 cultivated trees for Africa compared to 114 for China, but the list includes only one ficus and does not include marula, Berchemia discolor, baobab, and real fan palm); G. C. Kajembe, *Indigenous Management Systems as a Basis for Community Forestry in Tanzania: A Case Study of the Dodoma Urban and Lushoto Districts* (Wageningen: Wageningen Agricultural University Tropical Resource Management Papers, 1994); J. C. Okafor and A. Lamb, "Fruit Trees: Diversity and Conservation Efforts," in R. R. B. Leakey and A. C. Newton (eds.), *Tropical Trees*, pp. 34–41; J. R. Harlan, J. M. J. de West, and A. B. L. Stemler, "Plant Domestication and Indigenous African Agriculture," T. Shaw, "Early Crops in Africa: A Review of the Evidence," and J. W. Pursglove, "The Origins and Migrations of Crops in Tropical Africa," in Jack R. Harlan, J. M. J. de West, and A. B. L. Stemler (eds.), *Origins of African Plant Domestication*, pp. 3–19, 107–153, and 291–309, respectively. On *Faidherbia albida* (formerly *Acacia albida*), see J. Boutrais, *Mbozo-Wazan: Peul et Montagnards au Nord du Cameroun* (Paris, France: ORSTOM, 1987), pp. 45–48. On *Dacryodes edulis* see H. G. Adewusi, "Potential for Development and Conservation of *Dacryodes edulis* in Sakpoba Forest Reserve, Edo State, in the Niger Delta of Nigeria," T. Sunderland and O. Ndoye (eds.), *Forest Products, Livelihoods and Conservation: Case Studies of Non-Timber Forest Product Systems: Volume 2 Africa* (Jakarta, Indonesia: CIFOR, 2004), pp. 133–147. On *Parkia biglobosa*, see Smith et al., *Tropical Forests and Their Crops*, p. 425. Sunderland and Ndoye present case studies of forest products that they overwhelmingly depict as wild, but it is clear that most of the species discussed occur in or near farms and in secondary forest rather than in the wilderness. See T. Sunderland and O. Ndoye (eds.), *Forest Products, Livelihoods and Conservation: Case Studies of Non-Timber Forest Product Systems: Volume 2 Africa* (Jakarta, Indonesia: CIFOR, 2004). Okafor notes that the "forest plants" he discusses are, in fact, more numerous near settlements and stresses that trees retained in fields are relic vegetation; he refers to the trees in question as "semi-wild." See Okafor, "Case Study: Role of Conservation and Domestication of Minor Woody Forest Plants for Local Communities in Nigeria," pp. 7–17. Schoenbrun notes that the local name for the *Chlorophora excelsa* tree that marks many fields in the Central Lakes region of East Africa is the same word that is used for the hoe which he thinks links farming to the domesticated space represented by the tree. This suggests that the tree is not considered "wild" even though Schoenbrun does not imply in any way that the trees were planted or protected beyond the farmers having spared the tree when they cleared the farm lands. See Schoenbrun, *A Green Place*, p. 24. On tree planting by the Meru of Tanzania, see Neumann, *Imposing Wilderness*, p. 78.

[4] R. K. Hitchcock and J. I. Ebert, "Foraging and Food Production among the Kalahari Hunter/Gatherers," in J. D. Clark and S. A. Brandt (eds.), *From Hunters to Farmers: The*

American Indians cultivated oak, pine, mesquite, walnut, chestnut, willows, and palms using seeds or cuttings.[5] The notion that such trees are *natural* resources and are subject to *natural* resources management is a myth.[6]

This chapter builds on Leach and Fairhead's argument that forest islands in the savanna-forest mosaic of West Africa are not relic natural forests but human creations. But the chapter also moves beyond Leach and Fairhead's focus on inclinist outcomes to an analysis of the *processes* involved by demonstrating how, by whom, why, and in what context woody vegetation was changed.[7] The aim here is to highlight the extent to which the marula and other fruit trees that dominated the north-central Namibian landscape are neither entirely natural, nor wholly a human creation, but should instead be understood in terms of constituting environmental infrastructure. Using mutually exclusionist declinist or inclinist perspectives to assess environmental change in twentieth-century north-central Namibia creates a paradox because reforestation was preceded and accompanied by deforestation as refugees and settlers constructed new farms, fields, and villages in the floodplain's "wilderness" areas.[8] The paradox highlights the limitations of assessing environmental change in terms of deforestation and reforestation, terminology that is derived from and embedded in the Nature–Culture dichotomy. Instead of conceptualizing environmental change as a linear and singular process of degradation or improvement, or of stability or sustainability (and ultimately judging it as good or bad), the process of change should be reconceptualized in the plural: in terms of outcomes and baselines and

Causes and Consequences of Food Production in Africa (Berkeley: University of California Press, 1984), pp. 328–350. The authors mention "wild lemon" as being cultivated; they actually may be referring to the monkey orange (*Strychnos pungens*).

5 P. A. Williams, A. M. Gordon, H. E. Garnett, and L. Buck, "Agroforestry in North America and Its Role in Farming Systems," in A. M. Gordon and S. M. Newman (eds.), *Temperate Agroforestry Systems* (Wallingford, UK: Cab International, 1997), pp. 9–84 (especially 9–11).

6 Jackson and Chattopadhyay also reject the usefulness of the term "natural resources," but their main objection is to the *resources* part because the designation as resources implies that they only serve humanity. This is an important criticism. See C. Jackson and M. Chattopadhyay, "Identities and Livelihoods: Gender, Ethnicity, and Nature in a South Bihar Village," in A. Agrawal and K. Sivaramakrishnan (eds.), *Agrarian Environments*, pp. 147–169.

7 J. Fairhead and M. Leach, *Misreading the African Landscape: Society and Ecology in a Forest-Savanna Mosaic* (Cambridge: Cambridge University Press, 1996)

8 For a more detailed discussion of the deforestation of the middle floodplain, see Chapter 3.

trajectories of change, and consisting of a *series* of (sub)processes.[9] Environmental change was different on each farm, subject to the decisions, practices, ideas, and resources of each individual household (negotiated between its members) and each farm's unique soil and vegetation conditions. It was also subject to its own chronology. Typically, new villagers arrived at different times or in small groups. As a result, e.g., deforestation and reforestation processes on the various farms within a single village were out of sync.[10]

The actual day-to-day details of the process of cutting, propagating, using, and protecting woody vegetation and the who, the why, and the what have received little explicit attention.[11] A discussion of the dynamics of on-farm tree management, especially at the level of individual human actors, households, and individual trees, is critical. People and trees are more than simply a product of genetics. As with individual humans, individual trees each have a personal history and interactions result in

[9] Kreike, *Deforestation and Reforestation in Namibia*, pp. 3–4. Compare Jeffrey et al., who argue for a focus on multiple processes with different paths and subject to different forces with their own time-tables: "[i]ndeed, in human–environmental interactions it is often possible to observe the coupled development of two or more processes which could result in the collapse of an ecosystem (e.g., the sudden demise of a thriving population, or increased ecological vulnerability in the face of major anthropogenic perturbations)." See P. Jeffrey, M. Lemon, and B. Jefferson, "My Land, My Water, Your Problem: Co-dynamic Processes and the Development of Appropriate Water Policy Tools," in R. Coopey and T. Tvedt (eds.), *A History of Water, Volume 2*, pp. 529–549 (especially 536–543). Scoones emphasizes the need for plurality as well, distinguishing multiple pathways of environmental change, but envisages "multiple possible states . . . characterized by distinct physical, chemical and other features" instead of multiple processes. See Scoones (ed.), *Dynamics & Diversity*, pp. 31–36.

[10] For details on the processes in Ovamboland, see Kreike, *Re-Creating Eden*, pp. 35–186, and Kreike, *Deforestation and Reforestation*, pp. 23–46 and 177–195.

[11] On the paucity of research on this topic, see J. Clarke, W. Cavendish, and C. Coote, "Rural Households and Miombo Woodlands: Use, Value, and Management" and B. Campbell and N. Byron, "Miombo Woodlands and Rural Livelihoods: Options and Opportunities," in B. Campbell (ed.), *The Miombo in Transition: Woodlands and Welfare in Africa* (Bogor, Indonesia: CIFOR, 1996), pp. 101–135 (especially 134) and pp. 101–135 (especially 223), respectively. Crummey and Winter-Nelson provide selected narratives to show who planted what kind of trees but provide no details, e.g., about how they were planted or where, D. Crummey and A. Winter-Nelson, "Farmer Tree Planting in Wällo, Ethiopia," in T. J. Bassett, and D. Crummey (eds.), *African Savannas: Global Narratives and Local Knowledge of Environmental Change* (Oxford: James Currey, 2003), pp. 110–114. Kajembe provides lists of which trees were planted and stresses that most were planted to demarcate plots of land but does not specify who did the planting; see Kajembe, *Indigenous Management Systems*, pp. 99, 102–108. Fairhead and Leach argue convincingly that the village forests were ultimately created by human agency but focus on the reforestation outcome rather than on the processes involved. See Fairhead and Leach, *Misreading the African Landscape*.

unique characteristics.[12] Moreover, in Africa, individuals are responsible for most of the cut and planted trees, and each tree is individually cut or planted.[13] In addition, tree propagation in north-central Namibia displays a gendered pattern: women appear to have been more involved in on-farm fruit tree propagation and management. In contrast, men were more likely to be involved in removing, cutting, or coppicing trees and bushes, both on- and off-farm.[14] People propagated trees in north-central Namibia for

[12] On the importance of focusing on individual trees as the object of research, see Tomlinson and Zimmermann, Editorial Preface and D. H. Janzen, "Seeding Patterns of Tropical Trees," in P. B. Tomlinson and M. H. Zimmermann (eds.), *Tropical Trees as Living Systems* (Cambridge: Cambridge University Press, 1978), respectively, p. xv and pp. 83–128. On trees having histories, see P. Huxley, *Tropical Agroforestry* (Oxford: Blackwell Science, 1999), pp. 74, 174–197, and 241–242. Kozlowski, Kramer, and Pallardy, in *The Physiological Ecology of Woody Plants*, pp. xvii, 56, 81, 234, note that woody vegetation should be disaggregated to the level of individual trees. Moreover, they stress that because woody vegetation builds up and stores reserves over the years, its reserve levels at any given time are a reflection of their historical experience (i.e., droughts, pests, and herbivore browsing). See also F. Hallé, R. A. A. Oldeman, and P. B. Tomlinson, *Tropical Trees and Forests: An Architectural Analysis* (Berlin: Heidelberg, and New York: Springer Verlag, 1978), pp. viii–ix. Adaptation to different ecological conditions may lead to great intraspecies variation, e.g. – in terms of the onset of flowering – see M. Grouzis and M. Sicot, "A Method for the Phenological Study of Browse Population in the Sahel: The Influence of Some Ecological Factors," in H. N. Le Houérou (ed.), *Browse in Africa: The Current State of Knowledge* (Addis Ababa: ICLA, 1980), pp. 233–240. As a result, e.g., the palatability of browse to herbivores may greatly differ from plant to plant within a single species. See H.N. Le Houérou, "Browse in Northern Africa," in the same volume pp. 55–82, especially 61.

[13] Dejene found that farmers preferred to plant trees as individuals rather than through community projects, A. Dejene, *Environment, Famine, and Politics in Ethiopia: A View from the Village* (Boulder, Col.: Lynn Rienner, 1990), pp. 61–62. Crummey and Winter-Nelson highlight that tree planting was an individual undertaking; see Crummey and Winter-Nelson, "Farmer Tree Planting in Wällo, Ethiopia," pp. 110–114. Although they provide little detail, Franzel et al. also emphasize the farmer as an individual; see S. Franzel, P. Cooper, G. L. Denning, and D. Eade (eds.), *Development and Agroforestry: Scaling Up the Impacts of Research* (Oxford: Oxfam, 2002). See, e.g., the chapter by G. L. Denning, "Realising the Potential of Agroforestry: Integrating Research and Development to Achieve Greater Impact," pp. 1–14. Tree ownership by the planter also underlines that trees should be regarded as individuals. Bonnéhin, e.g., emphasized that the person who planted a tree remained its owner irrespective of the tree's location. See L. Bonnéhin, "Domestication paysanne des arbres fruitiers forestiers: Cas de *Caula edulis Baill. olacaceae*, et de *Tieghemella heckelii Pierre ex. A. Chev.*, sapotaceae, autour du Parc National de Tai, Côte d'Ivoire" (Wageningen Agricultural University: Ph.D. thesis, 2000), pp. 68–75.

[14] This chapter highlights on-farm tree propagation, an activity in which women played a major role. Chapters 3, 4, and 5 highlight on- and off-farm use and management of woody vegetation, including the clearing of farmplots and the regular cutting of main stems and branches (coppicing and pollarding), and the use of fire to shape tree and

diverse reasons: tree fruit, e.g., was a source of food, forage, and alcohol. Alcoholic beverages constituted a source of income especially for women, and the beverages were also critical capital for creating and maintaining social networks, which in turn were essential for security and for gaining access to land, labor, and other services and resources. A focus on the micro- and meso-levels of analysis (i.e., individual, household, and village) thus offers additional insights into the why and how of tree planting and cutting. But, rather than a priori celebrating north-central Namibians as natural environmental stewards or maligning them as natural environmental destroyers, this chapter argues that they were "architects of Nature," mentally and physically shaping the environment without being Nature's slave, master, or guardian.

Wilderness Gardens

The earliest European travelers to the Ovambo floodplain were struck by the abundance of fruit trees. In the early 1890s, upon entering the floodplain from the south, the missionary Wulfhorst noted the Ondonga kingdom's fertile fields, "here and there interspersed with groves of dark leafed trees or tall palm trees and through the leaves . . . large numbers of huts."[15] The "dark leafed trees" were most likely fig (*Ficus sycomorus*) trees, which grew to gigantic dimensions, reaching a height of eighty feet. The baobab attained the same height and its trunk grew to up to forty feet in circumference, but it was confined to the drier western half of the floodplain and the northern districts of Oukwanyama and adjacent

bush vegetation. Nygren points out that forest clearing is often seen as a male activity, A. Nygren, "Development Discourses and Peasant–Forest Relations: Natural Resource Utilization as Social Process," p. 25. Bonnéhin noted that both women and men planted trees but that men more commonly did so, Bonnéhin, "Domestication paysanne des arbres fruitiers forestiers," pp. 68–75. The gender dimension was considered to be of critical importance in agroforestry: men and women tended different trees and women typically were not integrated into agroforestry projects, see A. Böhringer, "Facilitating the Wider Use of Agroforestry for Development in Southern Africa," and M. D. Faminow, K. K. Klein, and Project Operating Unit, "On-Farm Testing and Dissemination of Agroforestry among Slash and Burn Farmers in Nagaland, India," in Franzel et al., *Development and Agroforestry*, pp. 35–55 and 84–106, respectively. On women and fruit tree planting, see also Tiffen, Mortimore, and Gichuki, *More People, Less Erosion*, p. 221.

15 A. Wülfhorst, *Aus den Anfangstagen der Ovambo-Mission* (Barmen: Verlag des Missionshauses, 1904), p. 4. See also B. Lau (ed.), *Carl Hugo Hahn Tagebücher 1837–1860; Diaries: A Missionary in Nama- and Damaraland* (Windhoek: Archives Services Division, 1985), part iv, p. 1039; NAN, NAO 104, Anderson to Hahn, extract diary Jordan, *Cape Quarterly Review*, 2 (1882), pp. 519–539.

Evale. Uukwaluthi in the southwestern floodplain boasted baobab and fig trees as well as palm trees.[16] Fig, real fan palm (*Hyphaene petersiana*), birdplum (*Berchemia discolor*), and marula (*Sclerocarya birrea*) occurred throughout the northern and southern floodplain, including Ongandjera, but the last two trees were not nearly as abundant elsewhere as in the northern floodplain. Huge jackalberry (*Diospyros mespiliformis*), birdplum, and marula dominated the northern floodplain landscape. Jackalberry lined the banks of the seasonal watercourses in the central and middle floodplain and the seasonal Etaka River in the southwest. During the 1880s and 1890s, missionaries waxed enthusiastic about the "wild" birdplum and marula trees and interviewees recalled that the trees had been abundant throughout pre-1915 Oukwanyama and Ombadja.[17]

During the last decades of the nineteenth century, palm wine was a popular beverage in the southern floodplain, but in the northern floodplain, marula tree and raisin bush fruit (*Grewia* spp.) were made into wines. In Oukwanyama, when the marula fruit ripened during the months of February and March, men and women engaged in celebrations for weeks on end and marula wine flowed freely. Not even recent mission converts could resist participating and a despairing missionary wrote: "[u]nfortunately [marula] trees are too abundant and everyone has them on their fields."[18]

16 Interviews by author: Johannes Shipunda, Omundaunghilo (Namibia), July 14, 1993; Kulaumoni Haifeke, Oshomukwiyu (Namibia), May 11, 1993; Elisabeth Ndemutela, Okongo (Namibia), February 16, 1993; Helemiah Hamutenya, Omuulu Weembaxu (Namibia), July 17, 1993; NAN, A233, J. Chapman, entries 1876 [pp. 61–62]; RCO 8 f. 9, Extract from RCO's Personal Diary, March 10, 1917, and RCO to Sec. SWA, Ondonga, October 27, 1918.

17 AVEM, RMG 2599 C/i 19, Bernsmann, Omburo, January 6, 1892; AGCSSp 466-A-VII, Duparquet, Carnets de Notes-Voyages # 3, entries August–September 1879 and info. Mr. Leen, April 29, 1879; Duparquet 1880 journal, July 10–11, 14, 1880. Interviews by author: Elisabeth Ndemutela, Okongo (Namibia), February 16, 1993; Helemiah Hamutenya, Omuulu Weembaxu (Namibia), July 17, 1993; Johannes Shipunda, Omudaunghilo (Namibia), July 14, 1993; and Julius Abraham, Olupito (Namibia), June 16, 1993. See also P. Möller, *Journey in Africa through Angola, Ovampoland and Damaraland* (Cape Town: C. Struik, 1974 [translated from the original Swedish edition of 1899]), pp. 125–126 and NAN, A233, J. Chapman, entries 1876 [pp. 61–62] and 1896 [146–147]; RCO 10, RCO, Notes for O/C Military Detachment Ovamboland, Ondonga, October 1916; A464, Archives FMS, Martii Rautanen Collection HP xxviii.1, Diary 1886–1888, Entry November 15, 1886 (the missionary Rautanen in the 1880s planted and transplanted fig saplings on his Ondangwa station).

18 NAN, NAO 104, Anderson to Hahn, diary Jordan; Möller, *Journey*, pp. 125. AGCSSp Duparquet, 1879 journal, August 17, September 10, 12, 14, 21, 23; C. Estermann, *Etnografia do Sudoeste de Angola: Vol. 1, Os Povos não-Bantos e o Grupo Étnico dos Ambós* ([Lisbon]: Ministério do Ultramar, Junta de Investicações do Ultramar, 1960), p. 134;

Despite their distaste for marula wine, however, the missionaries witnessed that tree fruit in the early 1900s proved to be invaluable during the seasonal period of shortages early in the rainy season (before the main field crops ripened) and during famine years.[19]

During the early 1900s, on-farm baobab trees in Omukwa village in the far northern floodplain were the property of the farm owners. When the fruit was ripe, the owners shared some of it with other households.[20] The marula tree was subject to complicated and overlapping rights. A portion of the marula wine that was produced had to be presented to the king of Oukwanyama. The households that occupied the land where the trees grew consumed the remainder.[21] In 1916–1917, British-South African troops who entered the floodplain to quell Oukwanyama resistance received strict orders to "[s]pare all fruittrees viz : Palm, Wild Fig, Marula etc. and all trees in inhabited areas" after the local colonial officials concluded that "[p]ractically all trees in such areas are wild fruit trees . . . and native owned."[22]

Fruit Trees and People

Colonial officials continued to be struck by the abundance of fruit trees in the northern floodplain in the second half of the 1920s. The Neutral Zone, a disputed area that encompassed the southern districts of Oukwanyama and was transferred to Portuguese sovereignty in 1927, was "exceptionally rich in wild fruits."[23] The inhabitants of the southern districts of Ombadja fled during the 1910s, a period that was marked by the violence of colonial conquest. In the late 1920s, the area was

Wülfhorst, *Anfangstagen*, p. 28; and quarterly reports in AVEM RMG 2518 C/h 34, Ondjiva, April 10, 1911, April 12, 1912, April 1, 1913, and June 30, 1914; RMG 2517 C/h 33 folio 10, Omupanda, April 1, 1914; RMG 2515 C/h 31, Omatemba, March 30, 1916.

19 AVEM, RMG 2599 C/i 19, Bernsmann, Omburo, January 6, 1892. On famines and fruit trees, see, e.g., RMG 2518 C/h 34, Ondjiva, April 1, 1913; Helena Nailonga, interview by author, Big Ekoka (Namibia), February 23, 1993; NAN, RCO 10 f. 15/1916/1, Major C.E. Fairlie, "Report of the Situation in the Ovakonyama Country," Namakunde, March 25, 1917.

20 Johannes Shipunda, interview by author, Omundaunghilo (Namibia), July 14, 1993.

21 AVEM, quarterly reports RMG 2517 C/h 33, Omupanda, April 1, 1914, and RMG 2518 C/h 34, Ondjiva, April 1, 1913.

22 NAN, RCO 10, "Notes for Officer Commanding Military Detachments Ovamboland, October 1916" and "Preliminary Memorandum for Expeditionary Force by Resident Commissioner C. N. Manning and Lt. Hahn [1916]."

23 NAN, A450 vol. 7, Annual Report Ovamboland 1926.

uninhabited, but an official noted "[m]uch of it was a perfect Paradise for natives abounding as it did in fruit trees and grazing."[24]

Monika Hidengwa was born in Ombuwa (in modern Angola), a location that lay outside of the settled zone of the precolonial Oukwanyama kingdom. Although trees and bush were plentiful, the area was devoid of birdplum, marula, and palm trees. During the 1920s, Monika Hidengwa moved with her family to Ombwabwa, near Omulunga (also in Angola). The only fruit trees in this village, which was also located beyond the settled heartland of Oukwanyama, were mangetti (*Ricinodendron rautanenii*).[25] An official who traveled east of Omafo into what was then considered uninhabited wilderness in the Angolan-Namibian border region noted a "singular absence of fruit trees which attract the natives in selecting settlement places."[26] Marula, birdplum, and palm trees began to appear after refugees from the northern Ovambo floodplain (in modern Angola) colonized the sparsely inhabited middle floodplain (along the border and south of the border in modern Namibia), clearing the bush vegetation, constructing farms, fences, and water holes, and laying out fields. Although these fruit trees were not entirely lacking in the middle floodplain before 1915, their occurrence was often associated with settlements that had been abandoned in the nineteenth century. For example, in the 1920s, when settlers from the northern floodplain arrived in middle floodplain Okalongo, they found palm and jackalberry trees, the relics of the early nineteenth-century kingdom of Haudanu: a cluster of palm trees at nearby Onandjaba was known as *Omilunga yaHaudanu*, or "the palm trees of Haudanu." The absence of contemporary settlement, however, meant that the region was considered to be wilderness and most of the fruit trees in the area's villages were propagated after the settlers had established their farms and fields.[27]

Only the southernmost district of the pre-1915 Oukwanyama kingdom, which included the village of Namakunde (in modern Angola),

[24] NAN, KAB 1, Draft Report Kunene Water Commission, Olusandja, July [1927], to Secretary SWA.

[25] Monika Hidengwa, interview by author, Eenhana Refugee Camp (Namibia), July 16, 1993. Ombuwa was probably a cattle post where her family sought refuge during the 1915 war.

[26] NAN, KAB 1, Submission to Administrator, Secretary and Attorney-General of SWA, 1927.

[27] Interviews by author: Julius Abraham, Olupito, June 15–16, 1993; Mathias Walaula, Onandjaba (Namibia), June 15, 1993; Kanime Hamyela, Omutwewondjaba (Namibia), June 15, 1993; personal observations, July 15, 1993; NAN, KAB 1, Volkmann, October 30, 1928, "Report on the Agricultural and Political Conditions."

stretched into the middle floodplain with isolated villages spilling over into what is Namibian territory today. In 1879, the missionary Duparquet noted clusters of fruit trees and signs of previous settlement at pans in this area, although at the time the sites were entirely abandoned. In 1915, most of the area was regarded as "wilderness."[28] Following the influx of tens of thousands of refugees resulting from the violent Portuguese conquest of the Oukwanyama kingdom in 1915, the landscape directly south of the colonial border changed dramatically as refugee-settlers cleared the expanses of bushland between isolated preexisting villages to create fields and construct new homes. A 1928 description of the new Oukwanyama district along the Angolan-Namibian border from west to east highlights the extent to which the landscape had been transformed: "[t]he first Ukuanyama werft [Oukwanyama homestead] is at Point [border marker] 16. The country becomes more open.... The werfts are on the dune-like banks; the bush has been destroyed. Everywhere one sees the beautiful Onjandi [*omwandi* or jackalberry] trees or wild figs."[29]

The impact of settlement on vegetation cover was especially dramatic in the eastern half of the middle floodplain, where a new Oukwanyama district had been carved out of sparsely inhabited bush lands in less than fifteen years. In 1931, the Assistant Native Commissioner for Oukwanyama, newly arrived at his post, was shocked by the "[m]any cases of wilful [*sic*] destruction of large, often fruit, trees... to quote the words of Mr. Hahn in one particular case, '[t]he area had altered so much, due to destruction of trees, that it was hard to believe it was the same area.'" The eager official immediately ordered the district's headmen to reissue and strictly enforce "the old tribal law... protecting all fruit trees, palm trees and trees, the trunk of which is thicker than the thigh of a man." Native Commissioner Hahn, however, played down his subaltern's alarm by stressing that tree clearing for new farms was unavoidable. Hahn reassured his superior that the chiefs and headmen generally protected fruit trees in the "tribal" districts.[30]

28 AGCSSp Duparquet, 1879 journal, August–September 1879; Malita Kalomo, interview by author, Omutwewondjaba (Namibia), July 15, 1993.

29 NAN, KAB 1, Volkmann, October 30, 1928, "Report on the Agricultural and Political Conditions." On the violence and displacement caused by the imposition of colonial rule during World War I, and the subsequent massive woody vegetation clearing, see Kreike, *Deforestation and Reforestation*, pp. 23–43.

30 NAN, NAO 44, O/C Oshikango to NCO, Oshikango, March 17, 1931, and NCO to Sec. SWA, Ondangwa, April 20, 1931.

The first two heads of the colonial administration in Ovamboland, Hahn and Eedes thought it was unnecessary to introduce or enforce South Africa's tree conservation regulations in the area.[31] Only in 1954, after Hahn's successor Eedes retired, and the Native Affairs Department of the Union of South Africa took control over the reserves of colonial Namibia, were South African conservation policies introduced lock-stock-and-barrel in north-central Namibia. Although Ovamboland's agricultural officer concluded in 1957 that fruit trees were already being protected by the local population, he nevertheless recommended compiling a list of "valuable trees" that could not be cut down and distributing the list to all the district and village headmen. In 1978, a newly appointed forester for Ovamboland, apparently unaware of these earlier attempts, compiled a separate list of protected trees that included all the principal indigenous fruit trees. He, too, however, acknowledged that most actually were protected under "Owambo [Ovambo] tradition."[32]

Colonial conservation measures had an impact. Interviewees recalled that fully grown indigenous fruit trees could not be cut down without the permission of the village and the district headmen, even if they were located on a household's farmlands. Philippus Haidima mentioned that cutting down a marula tree could even get one arrested. Permission to cut down a fruit tree was granted only when, e.g., the tree contained a hole that housed snakes. The restriction also applied to the species' non-fruit-bearing male trees, which previously had enjoyed little protection. Cutting the branches of indigenous fruit trees was restricted.[33]

The colonial government also attempted to limit tapping palm juice from the trunks of adult trees (which could kill the tree) and championed "customary" preservation regulations to prevent fruit trees from being cut down in the inhabited areas of the floodplain.[34] In 1916, the South

[31] NAN, NAO 44, Officer Oshikango to NCO, Oshikango, March 17, 1931, and NCO to Secretary SWA, Ondangwa, April 20, 1931.

[32] NAN, NAO 61, Quarterly Reports, April–June and July–September 1954; BAC 133, Agricultural Officer to CNC, Ondangwa, November 19, 1956, and to NCO, Ondangwa, July 4, 1956; BAC 131, Agricultural Officer to Bantu Commissioners Ondangwa and Oshikango [Ondangwa], January 28, 1957; OVJ 19, Forester Owambo to Secretary, [Ondangwa], August 18, 1978. See also A. Erkkilä and H. Siiskonen, *Forestry in Namibia, 1850-1990* (Joensuu: University of Joensuu, 1992), pp. 74–77.

[33] Interviews by author: Kaulikalelwa Oshitina Muhonghwo, Ondaanya (Namibia), February 2, 1993; Philippus Haidima, Odibo (Namibia), December 9, 1992; Joseph Nghudika, Onamahoka (Namibia), February 3, 1993.

[34] NAN, NAO 101, NCO to CNC, Ondangwa, April 17, 1948; NAO 44, NCO to CNC, Ondangwa, June 2, 1941; NAO 43, NCO to CNC, Ondangwa, August 26, 1946, and telegram NCO to Secretary SWA, Ondangwa August 26,1946; NAO 60–61, Quarterly

Africans issued a proclamation in Ovamboland that included the stipulation that no palm trees could be damaged.[35] During the famine years of the late 1920s and the early 1930s, however, desperate individuals tapped palm juice on such a scale that it contributed to the destruction of large numbers of palm trees and deforestation in the southern floodplain.[36] In the 1950s, fines for illegal palm juice tapping were draconian: in Ombalantu district the fine was five head of cattle or 25 pounds several times more than the amount of bride wealth or the fee for a farm.[37] In 1972, however, the Director of Agriculture for Ovamboland sought to reassure the Secretary for Bantu Administration in Pretoria that his fears that palm trees were declining in number were groundless.[38]

The Fruit Tree Frontier

The introduction of marula and birdplum trees in the middle floodplain accompanied settlement. In early 1930s Omupanda in modern Namibia, marula trees occurred only on two farms, including the first farm that had been established in the area three decades previously.[39] Although the western Ombalantu district had a good number of marula trees and even some birdplums in the late 1960s and the central Uukwambi district had also acquired those trees by the early 1990s, many of the trees had been brought by settlers when they established new villages in the "wilderness" stretches that had separated the preconquest Ovambo polities. The seeds themselves may have originated from the middle floodplain or further north in Angola.[40]

Reports October–December 1952 and April–June 1954; BAC 133, Agricultural Report Ovamboland 1956/1957. On forestry, see Erkkilä and Siiskonen, *Forestry in Namibia*, pp. 65–80.

35 NAN, RCO 4, "Proclamations by Chiefs, Notice: Orders to Headmen and People."

36 NAN, NAO 18–19, Monthly Reports, November–December 1928, January–February 1929, April 1930, March 1931, July–August 1932.

37 NAN, NAO 51, Proceedings of an Enquiry held at Ombalantu on June 15, 1954, in the Allegation against Headman Hishitile Shiweda. See also NAO 71 f. 32/3, Chief Kambonde to NCO, Okaloko, December 2, 1953, and NCO to Kambonde, December 4, 1953.

38 NAN, OVA 49, Sec. Bantu Administration to Director Agriculture, Pretoria, February 9, 1972; State Forester to Director Agriculture, Ondangwa, February 21, 1972; Director Agriculture to Sec. Bantu Administration, Ondangwa, February 25, 1972.

39 Interviews by author: Mateus Nangobe, Omupanda (Namibia), May 24, 1993; Paulus Wanakashimba, Odimbo (Namibia), February 10–11, 1993; Paulus Nandenga, Oshomukwiyu (Namibia), April 28, 1993; Kulaumoni Haifeke, Oshomukwiyu (Namibia), May 11, 1993.

40 On Uukwambi, personal Communication with Joseph Hailwa, Regional Forester Ovamboland, March 24, 1992. On the occurrence of marula and birdplum in late 1960s

Fruit trees were not only associated with human settlement; human action also caused their southward expansion, although the extent to which people intervened in "natural" processes varied. Kulaumoni Haifeke, the 1930-born daughter of the first pioneers to settle Oshomukwiyu, saw her village change from an uninhabited "wilderness" into a landscape filled with villages and farms rich in full-grown marula, birdplum, and palm trees and concluded: "only God makes them grow."[41] Others highlighted the role of humans, and women in particular, in tree propagation; women, e.g., imported the fruit from the northern floodplain.[42] The health officer of Ovamboland noted in 1933 that the pits were often consumed along with the fruit of the birdplum. He counted twenty-six birdplum pits in a single stool of a small child and surmised that this was the reason that birdplum seedlings could be observed everywhere around the homesteads.[43] Mathias Walaula stressed that in Onandjaba, fire was used to thin the palm bushes because only a free standing palm bush could grow into a tree, and Kanime Hamyela proudly recounted how he had shaped the dense bush into a fertile garden with stately trees: "[t]he plants are like grains. If you thin millet it will grow fast and properly. It is the same with the bushes – if you cut out some then the remaining will grow fast and healthy."[44] In 1993, the large and shady marula tree on Julius Abraham's Olupito farm produced enough fruit to brew a thirty-liter pot of marula wine that he used to entertain his friends and guests. His father – one of the pioneers to settle Olupito – had discovered the tree when it was a mere bush hemmed in by huge trees. After felling the trees around the bush, his father kept the surroundings free from weed. In addition, one of the older villagers, Joseph Kashinghola, had watered his own birdplum treelings and over time they had developed into impressive full-grown trees.[45] Pauline commented that during her childhood: "[a]ll people took care of these trees.... Young men of today don't take care of plants (trees); they cut them down or plow

Ombalantu, see NAN, OVA 57, Lueckhoff, Report on Visit to SWA, November 3–15, 1969, Appendix to Regional Forester to Chief Director Bantu Administration, Grootfontein, April 3, 1970.

41 Kulaumoni Haifeke, interview by the author, Oshomukwiyu (Namibia), May 11, 1993.

42 Paulus Nandenga, interview by the author, Oshomukwiyu (Namibia), April 28, 1993.

43 Paulus Wanakashimba, interview by the author, Odimbo (Namibia), February 10–11, 1993; NAN, NAO 36, Annual Health Report Ovamboland 1933.

44 Interviews by author: Mathias Walaula, Onandjaba (Namibia), June 15, 1993, and Paulus Wanakashimba, Odimbo (Namibia), February 10–11, 1993; Kanime Hamyela, interview by the author, Omutwewondjaba (Namibia), June 15, 1993.

45 Julius Abraham, interview by the author, Olupito (Namibia), June 16, 1993.

them under.... Because they were not abundant we took care of every birdplum, jackalberry, and fig [tree]."[46]

Settlers carried fruit, cuttings, and seeds from the northern and middle floodplain into eastern Ovamboland. Pioneers entered the area from the floodplain in the 1920s. Whereas marula and other fruit trees were still scarce in the small eastern villages in the 1930s, by the 1990s birdplum trees could be found on fields east of Okongo, although they were rare toward the Okavango region.[47]

Interviewees from the floodplain and from the most western part of eastern Ovamboland tended to stress that the fruit trees that appeared in their new villages and on their farms grew "naturally" or were "gifts of God," even if the seedlings had been carefully nurtured. Interviewees from the far eastern Ovamboland, however, tended to emphasize human agency in fruit tree propagation.[48] Eastern Ovamboland was one of the most recently settled areas of the region and had been virtually devoid of villages (and the principal fruit trees) before the first pioneers arrived from the floodplain. Perhaps as a result, the introduction of fruit trees was more explicitly associated with human action.

Women especially valued fruit trees because the food source could be consumed fresh, or it could be dried and stored. In addition, alcoholic beverages that were fermented or distilled from fruit were critical means to maintaining social security and patronage networks. Finally, the sale of alcoholic beverages, distilled liquors in particular, was the sole means by which women could earn cash.[49] Although the colonial administration declared home distilling illegal, the economic crisis of the late 1920s and early 1930s caused a severe decline in the demand for migrant laborers from north-central Namibia, and for many households the sale

[46] Pauline, interview by author. Onenghali (Namibia), December 15, 1992.

[47] NAN, A450 vol. 7, Annual Report 1935; Interviews by the author: Kalolina Naholo, Ohamwaala (Namibia), January 26–27, 1993; Paulus Nandenga, Oshomukwiyu (Namibia), April 28, 1993; Timotheus Nakale, interview by the author, Ekoka laKula (Namibia), February 21, 1993; Franscina Herman, Odibo (Namibia), December 12, 1992; Kaulikalelwa Oshitina Muhonghwo, Ondaanya (Namibia), February 2, 1993; Moses Kakoto, Okongo (Namibia), February 17, 1993; Helemiah Hamutenya, Omuulu Weembaxu (Namibia), July 17, 1993. On the colonization of eastern Ovamboland by floodplain settlers, see Kreike, *Recreating Eden*, pp. 129–154.

[48] Interviews by the author: Alpheus Hamundja, Ohamwaala (Namibia), January 26, 1993; Joseph Nghudika, Onamahoka (Namibia), February 3, 1993; Nahandjo Hailonga, Onamahoka (Namibia), February 4, 1993; Kaulikalelwa Oshitina Muhonghwo, Ondaanya (Namibia), February 2, 1993.

[49] For details about the uses of various fruits, see NAN, BAC 133, Agricultural Report Ovamboland 1956/1957.

of liquor distilled from birdplum fruit provided an alternative source of cash income.[50]

Ovamboland's Oukwanyama and Ondonga districts, as well as the Lower Kunene province of Angola, were the principal liquor production centers.[51] The importance of liquor distilling is underlined by the long and failed crusades against home distilling that the colonial administration waged from the 1930s to the 1950s. Colonial staff destroyed more than 400 stills in north-central Namibia during 1947–1949 alone. Forty years later, however, more than 40 percent of the respondents to the 1993 OMITI survey reported that they sold homemade liquor. As an income generating enterprise, home distilling was topped only by the sale of handmade Ovambo baskets.[52]

Although they were not important for distilling, exotic fruit trees also played a role in the expanding fruit tree frontier from the 1940s onward. The 1940 annual report for Ovamboland noted that papaya cultivation was rising, especially in Oukwanyama district. A 1945/1946 colonial report estimated that there were approximately 100 citrus trees in north-central Namibia, a figure that probably understates the spread of exotic fruit trees in Ovamboland's villages at the time.[53] Fruit trees, including papaya, guava, and lemons, could sometimes be found in the 300 horticultural gardens that colonial officials had identified in north-central Namibia. Vegetables, including tomatoes and onions, and tobacco were the main horticultural crops. The majority of such gardens were located in the Oukwanyama district, especially around Endola. The estimated 2,000 exotic fruit trees (papaya and guava) in north-central Namibia in the mid-1950s mostly were thought to be located in the Oukwanyama district

[50] NAN, NAO 18, Monthly Report December 1928. NAO 11, FMS to NAO, Olukonda, November 12, 1928, and NAO to FMS, Ondangwa November 25, 1928; interviews by author: Mwulifundja Haiyaka, Omhedi (Namibia), March 8, 1993; Franscina Herman, Odibo (Namibia), December 12, 1992; and Philippus Haidima, Odibo (Namibia), December 9, 1992.

[51] NAN, NAO 18, Monthly Report December 1928. NAO 11, FMS to NAO, Olukonda, November 12, 1928, and NAO to FMS, Ondangwa November 25, 1928, and interviews by author: Mwulifundja Haiyaka, Omhedi (Namibia), March 8, 1993; Franscina Herman, Odibo (Namibia), December 12, 1992; and Philippus Haidima, Odibo (Namibia), December 9, 1992.

[52] NAN, A450, vol. 12, SWA Commission, vol. 12, pp. 671–672; NAO 71, NCO to ANC, Ondangwa, May 12, 1947; NCO [Ondangwa], August 29, 1947; NCO to CNC, Ondangwa, September 14, 1947; ANC to NCO, Oshikango, November 13, 1947; NAO 71, NCO to Chief Kambonde, Ondangwa, December 28, 1948 and to CNC, Ondangwa, February 20, 1949; BAC 133, Agricultural Report Ovamboland 1956/1957; OMITI 7.2.

[53] NAN, NAO 103, Census of Agriculture Ovamboland 1945/1946.

and "in general . . . are randomly planted next to the kraals." Sometimes, such fruit trees were hand-watered as was the case with other horticultural crops. As with horticulture in general, however, water scarcity was a major constraint: "[f]ruit cultivation [occurs] on very small scale because water is scarce and at certain times in fact only for domestic use."[54]

Trials at the colonial Mahanene nursery in the late 1970s included mango and papaya; and, by the late 1970s, the Department of Agriculture was marketing seedlings grown under irrigation (probably at Mahanene). In 1979/1980, the Department sold 200 papaya and 20–30 other treelings, including citrus plants.[55] In the early 1990s, the annual production of the Danish Aid from People to People (DAPP) tree nursery at Ombalantu, one of less than a handful of tree nurseries in north-central Namibia, was 3,000 guava, 1,000 jacaranda, 100 Acacia, 2,500 *Leucaena leucocephala*, 800 Eucalyptus, and 100–200 other species. DAPP planted the trees depending on demand and popularity.[56] DAPP income decreased in the dry season because "many people buy during rainfall where water is available and plants grow fast because they get enough water. During [the] dry season there is no [*sic*: not] enough water for watering trees."[57]

The occurrence and spread of indigenous and exotic fruit trees alike was thus closely associated with human agency. They were located on-farm and respondents to the 1993 OMITI survey emphasized the intense management of fruit trees that involved protecting the trees with palisades and fences. According to the survey, marula and birdplum were by far the most frequently seeded local trees. Marula was also actively propagated by cuttings. Thirty-nine and 48 percent of the respondents reported that they protected marula and birdplum treelings, respectively. Respondents to the 1993 OMITI survey ranked indigenous fruit trees

54 On gardens and fruit trees, see NAN, A450 vol. 7, Annual Report Ovamboland 1940; BAC 133, Agricultural Reports Ovamboland 1955/1956 and 1956/1957 and Quarterly Report Agriculture Ovamboland, June 30, 1958. See also AHE (BAC) 1/352, Annual Report for Agriculture Ovamboland 1968.

55 NAN, OVA 50, Sec. Agriculture to Head Research Institute for Citrus and Subtropical Fruits Nelspruit, Ondangwa, August 14, 1974; OVA 6, Annual Report Agriculture Owambo 1979/1980; Jerry Ileka (Agricultural Officer Mahanene Research Station), interview by author, Mahanene (Namibia), March 20, 1992.

56 Sheuyange Tufaneni (second-year student, Agriculture, Ogongo Agricultural College), report of a field trip to the DAPP nursery, Ombalantu, n.d. [July 1993].

57 Saara Nehale (second-year student, Agriculture, at the Ogongo Agricultural College), report field trip to DAPP nursery, Ombalantu, n.d. [July 1993]. Three nurseries operated in Ovamboland in 1993: the government nursery at Ondangwa, the Rural Development Center at Ongwediva (since 1988), and DAPP; see Erkkilä and Siiskonen, *Forestry in Namibia*, pp. 135–136.

as the most important trees in their villages. Birdplum and marula were each mentioned by more than 60 percent of the respondents; palm by almost half; and jackalberry by almost one-third. When queried about the trees that they did not already have on their farms but would like to add, respondents mentioned a mixture of exotic and indigenous trees, principally fruit trees (notably guava, citrus, birdplum, and marula).[58]

Farms and Fields as Tree Nurseries

To colonial officials, the location of fruit trees in the landscape appeared random, an impression that reinforced their misperception that indigenous fruit trees were "wild" and "wilderness" trees. In his report for 1956–1957, the agricultural officer stated that the abundant indigenous "natural fruit tree species ... grow without any care and succeed well in meeting the needs of the population.... [T]hey really occur everywhere in the forested areas of the region."[59] The "forested areas" to which the agricultural officer referred were located in the densely inhabited parts of Ovamboland, however, and *not* in the uninhabited "wilderness" stretches. Moreover, the fruit trees were located inside the villages, on farms and fields, and not in the bush on the margins of the villages or between farms.

The actual homestead, consisting of living quarters, kitchens, and storage huts and surrounded by a palisade, was the perfect nursery for the active and passive propagation of fruit trees (also, ironically, because most of the competing tree vegetation had been removed). Fruit was prepared and eaten within the palisaded homestead; pits and used water were discarded around the huts or on adjacent household middens. Under favorable conditions, pits sprouted and developed into treelings. According to the 1993 OMITI survey, 41 percent of the households surveyed had engaged in seeding trees, mostly since the mid-1950s; 90 percent had planted the seeds within the homestead; and *all* of the households had seeded the trees on-farm. Tree propagation in other forms also centered

58 OMITI 4.4.29 (*N*=353) and 4.4.43 (*N*=325). It may also be significant that jackalberry and marula were listed as "sacred trees" by a relatively high percentage of the respondents. The sample was small (*N*=49), but jackalberry ranked first (mentioned by 20%) with marula in a shared fifth place (8%), and with baobab and birdplum mentioned as "sacred" by 4% each. Birdplum was not listed among the possible answers on the questionnaire; the respondents volunteered it.

59 NAN, BAC 133, Agricultural Report Ovamboland 1956/1957. See also NAO 17, ANC to NCO, Oshikango, January 25, 1942; BAC 123, Famine Relief Schemes Oshikango vol. 2, Bantu Affairs Commissioner Oshikango to Chief Bantu Affairs Commissioner, Oshikango, November 15, 1961.

on the homestead. Cuttings were planted in the homestead by 86 percent of the respondents; 89 percent of the respondents reported (re)planting treelings in the homestead; and in all other cases, treelings were propagated mainly on-farm. Half of the OMITI survey respondents stressed that they protected naturally occurring trees within the palisaded homestead itself, and half again reported doing so within the confines of the farm fence that surrounded the homestead, fields, and fallow. In contrast, only a few respondents mentioned protecting naturally occurring trees off-farm, and, in all cases, the concerned trees were located just outside the farm fence. The palisade that marked the homestead offered saplings protection from livestock and from the seasonal ravages of the burning sun, merciless winds, and frost. The entire palisaded homestead was generally relocated within the farm every three to five years; by then, treelings were much better prepared to face the elements and livestock. Full-grown fruit trees therefore often marked former locations of the enclosed homestead.[60]

[60] Interviews by the author: Kaulikalelwa Oshitina Muhonghwo, Ondaanya (Namibia), February 2, 1993; Moses Kakoto, Okongo (Namibia), February 17, 1993; Timotheus Nakale, Ekoka laKula (Namibia), February 21, 1993; Helemiah Hamutenya, Omuulu Weembaxu (Namibia), July 17, 1993; Philippus Haidima, Odibo (Namibia), December 9, 1992; Paulus Wanakashimba, Odimbo (Namibia), February 10–11, 1993; OMITI 4.4.1–3, 4.4.7, 4.4.11, 4.4.16. In his overview of agroforestry projects in Africa, Kerkhof noted that few projects conducted initial surveys but stresses that in the cases where project staff did undertake research, they were surprised to learn that farmers commonly planted trees, P. Kerkhof, *Agroforestry in Africa: A Survey of Project Experience* (London: The Panos Institute, 1990), pp. 52–60, 81, 115, 143–148, 168. Kajembe observed that indigenous trees were disappearing from crop fields and that they were concentrated in home gardens or used to mark plot boundaries, see Kajembe, *Indigenous Management Systems*, pp. 99, 113. Bonnéhin stressed that the *Tieghemella heckelii* tree that began to be domesticated in the late 1960s was often found around the locations of old settlements or camps; see Bonnéhin, "Domestication paysanne des arbres fruitiers forestiers," pp. 68–69. Kessy noted that some trees had been domesticated on-farm in East Usambara. He surveyed eighteen home gardens to identify "domesticated species" but provides no details concerning how and the extent to which these trees had been domesticated; see Kessy, *Conservation and Utilization of Natural Resources*, pp. 88, 104–106. In his study area in the western part of eastern Ovamboland, Erkkilä observed that the fruit trees that were located in the center of the fields had the highest crown density and surmised that the center of the field indicated the oldest cultivated section. It is probable, however, that the location marked the location of the original palisaded homestead. In addition, Erkkilä points to a farmer in the Omusati region (in the floodplain) who stressed that he had moved his homestead because marula trees had begun to grow in it. Erkkilä also identifies the fig tree as an introduced tree in Ondobe. See A. Erkkilä, "Living on the Land: Change in Forest Cover in North-Central Namibia, 1943–1996" (PhD diss., University of Joensuu, 2001), pp. 92, 97–98. Verlinden et al. emphasize that fruit trees in Ovamboland are associated with farms but claim that most were sown "casually," suggesting that these were not really domesticated; see A. Verlinden, M. K. Seely,

Colonial officials' observations indirectly confirm that the most important fruit trees – palm, marula, birdplum, fig, and baobab – were located in the village landscapes of Ovamboland, as they were often explicitly described as being associated with habitation and homesteads. A manuscript written by the Native Commissioner during the 1920s or 1930s stated: "[i]n the inhabited areas are to be found various wild fruit trees.... " In a 1942 letter, the Native Commissioner emphasized that "wild fruits" were part of the diet but that these sources of food were located near the homesteads and "[were], however, not available when working in the bush." In 1953, elephants damaged a large number of "*home-trees* [emphasis author] especially... marura [marula] and palm trees – necessary trees which supply people with nutritive food." In 1957, the agricultural officer for Ovamboland wrote "[t]ree species that carry fruit such as Maroela [marula], Wildevy [fig], Jakkalvrug [jackalberry]... usually are not eradicated because of their economic and food value. The Makalanie [real fan] palm is also protected because of its fruits." In 1961, the Deputy Secretary for Forestry of the Union of South Africa followed the border road from the Okavango region in the east to Oshikango in the west and noted that the landscape changed from a more or less closed forest beginning in the Okavango region to twenty to twenty-five miles east of Oshikango, to a much more open landscape west of this point, with beautiful marula, jackalberry, and fig trees scattered in the Ovambos' fields.[61] The 1963 Odendaal Commission reported: "[w]estwards towards the oshana region of Ovamboland [i.e. the floodplain] the bush becomes sparser and the ana tree (*Faidherbia albida*), mopane and palm... make their appearance. Marula and manketi [mangetti], as well as wild fig and other kinds of trees such as omwaandi [jackalberry] also occur here.... In the southern region of Ovamboland the palm belt merges into extensive grassy plains."[62] Another 1960s description reads, "[t]he first impression of Ovamboland is trees, trees, and more trees. Makalani palms, Mangetti, Maroella

and A. Hillyer, "Settlement, Trees, and Termites in Central North Namibia: A Case of Indigenous Resource Management," *Journal of Arid Environments*, 66 (2006), pp. 307–335.

61 NAN, A450 vol. 10, "Tribal Affairs"; NAO 17, ANC to NCO, Oshikango, January 25, 1942; NAO 67, Tribal Secretary Uukwaluthi to NCO, Uukwaluuthi, September 22, 1953; BAC 131, Agricultural Officer Ovamboland to Bantu Commissioners Ondangwa and Oshikango, January 28, 1957, and BAC 131, Deputy Secretary of Forestry, "Report of a visit by the Deputy Secretary of Forestry to South West Africa: April 17–29, 1961," Pretoria, May 10, 1961.

62 Report of the Commission of Enquiry into SWA Affairs 1962–1963, December 12, 1963, p. 9.

[marula] trees and the wild Fig tree and many others."[63] Despite the evidence that the bulk of the fruit trees occurred inside the villages, however, colonial officials did not revisit the overlapping assumptions that the fruit trees were "wild" and that they were the sole relics of a previous "natural" forest cover.

Maintaining Select Trees in Fields

Although land clearing destroyed large amounts of woody vegetation, pioneer farmers spared a few selected trees.[64] When Timotheus Nakale's father cleared a new farm plot at Onakalunga in the 1940s, he preserved a monkey orange (*Strychnos cocculoides*) tree for its fruit and a couple of shepherd's trees (*Boscia albitrunca*) to provide his family with a shady place to rest while they were cultivating.[65]

In the early 1990s, surveyed and interviewed farmers expressed a clear awareness about the advantages and disadvantages of having on-farm trees. Trees functioned as windbreaks. This was especially necessary in far eastern Ovamboland (the eastern part of modern Ohangwena) where strong winds carry sand and dust particles that have the potential to damage crops on impact, especially during the vulnerable flowering stage. Particles could also cover plants with a fine layer of dust that interferes with photosynthesis.[66] Moreover, trees in and around fields reduce water loss through evaporation and evapotranspiration and trees can buffer temperature extremes (frost occurs occasionally in eastern Ovamboland).

Farmers also demonstrated a significant awareness about tree-soil relationships. In 1957, the agricultural officer for Ovamboland wrote: "[i]n many cases densely grown parts are consciously cut and burned to make room for fields, because those areas are much more fertile as [*sic*] the parts without trees . . . or those [parts] that are only covered by . . . inferior

[63] NAN, WAT 3 f. 17, S. Davis, "Tour of Northern Territories – Some Random Observations, Comments, and Thoughts [approximately 1960]."

[64] NAN, NAO 104 Anderson to Hahn, extract diary Jordan, pp. 519–539; A450, 9, "Property Rights;" and BAC 131, Agricultural Officer Ovamboland to Bantu Commissioners Ondangwa and Oshikango, January 28, 1957. See also interviews by the author: Julius Abraham, Olupito (Namibia), June 16 and 18, 1993; Timotheus Nakale, Big Ekoka (Namibia), February 21, 1993; Mateus Nangobe, Omupanda (Namibia), May 24, 1993.

[65] Timotheus Nakale, interview by the author, Big Ekoka (Namibia), February 21, 1993. See also NAN, BAC 131, Agricultural Officer Ovamboland to Bantu Commissioners Ondangwa and Oshikango, January 28, 1957.

[66] Interviews by the author: Timotheus Nakale, Big Ekoka (Namibia), February 21, 1993, and Joseph Kambangula, Omboloka (Namibia), February 25, 1993; and OMITI A5.2.3.

bushes."[67] Trees marked certain soils and certain soils were marked by particular trees or bush. The clayish *omutunda* soil, e.g., supported mopane, tamboti, and red bushwillow (*Combretum apiculatum*) trees and sandpaper raisin (*Grewia flavescens*) bush, whereas the sandy *ehenene* or *omahenene* soil supported mainly mopane bush. *Omufitu* soil supported birdplum on the side of the watercourses, as well as wild seringa and Transvaal teak.[68] In addition, farmers acknowledged that trees affected and changed soil composition for the better or for the worse. Farmers hailed shepherd's tree, (Kalahari) apple-leaf, African wattle (*Peltophorum africanum*), and marula, e.g., for their beneficial effects on the soil.[69]

Farmers carefully weighed both the advantages and disadvantages of having on-farm trees in general and specific species in particular. In 1993, among reasons given to explain the advantages of on-farm trees, close to 80 percent of the respondents mentioned shade, ranked in frequency only after the importance of trees as sources of fruit and fodder, but before firewood. Sixty percent of the respondents mentioned esthetic reasons, ranked after firewood. Forty-four percent of the respondents claimed that on-farm trees were beneficial for crops; 42 percent of the respondents mentioned that trees were beneficial for the soil; and 35 percent of the respondents explained that trees reduced water loss.[70]

In explaining the disadvantages of on-farm trees, 77 percent of the respondents mentioned that trees inhibited plowing. The trunks and roots of felled trees routinely were left in the fields, which made plowing difficult. Plowing also is more likely to damage small trees that are not yet established in the fields. Fig, mangetti, and Transvaal teak trees were considered to render the soil useless for crop cultivation because of their "bad roots," as a farmer from eastern Ovamboland (Ohangwena region) put it.[71] Forty-six percent of the respondents mentioned that trees drained the soil of nutrients and 44 percent of the respondents emphasized that trees were bad for crops.[72]

67 NAN, BAC 131, Agricultural Officer to Bantu Commissioners Ondangwa and Oshikango, January 28, 1957.

68 Abisai Dula, interview by the author, Oikokola (Namibia), June 25, 1993. *Ehenge* in the Kwanyama dialect is the equivalent of *mahenene* or *henene* in the Ndonga dialect.

69 OMITI 4.4.26 (N=354).

70 OMITI 4.4.26 (N=354).

71 Timotheus Nakale, interview by author, Ekoka laKula (Namibia), February 21, 1993.

72 OMITI 4.4.27. For this question, N=144; i.e., half of the universe for 4.4.26, which suggests that half of the sample was of the opinion that on-farm trees had no disadvantages.

Specific trees could also harbor birds that could be crop pests, and bird scaring and hunting were critical.[73] In fact, almost three quarters of OMITI households considered birds to be a threat to their crops.[74] On balance, however, the large majority of the region's inhabitants valued selected on-farm trees and preserved and often propagated them because trees constituted important environmental infrastructure.

Conclusion

Although such indigenous fruit trees as marula abounded in northern Ovambo floodplain villages in the early twentieth century, they were rare in the largely uninhabited middle floodplain and in the southern floodplain until they were introduced by refugees from the northern floodplain. The resulting process of reforestation in Ovamboland appears to sharply contradict studies that have argued that north-central Namibia's twentieth-century history is marked by dramatic deforestation.[75] The marula, birdplum, and palm trees that dot north-central Namibia's fields are neither "wilderness" nor "wild" trees and thus do not constitute the relic vegetation of an earlier extensive forest cover.[76] Instead, they were propagated both passively and actively through human agency (e.g., via seeds and cuttings), and they were as rare in the uninhabited "wilderness"

73 On birds as a pest and/or trees and birds, see interviews by the author: Timotheus Nakale, Ekoka laKula (Namibia), February 21, 1993, and Marcus Paulus, Olukula (Namibia), February 19, 1993; NAN, NAO 19, Monthly Reports Ovamboland September–October 1931; NAO 45, statement Andreas Zacharias, Onayena, January 18, 1940; NAO 51, Kaundjua Andunge to NCO, Ombalantu, April 16, 1950; NAO 91, statement Titus Muatelai Kakonda, Ondangwa, May 10, 1954; OVA 47, A. C. Venn to Sec. Agriculture Owambo, March 12, 1976; OVA 58, Forester to Principal Ongwediva, October 15, 1979.

74 OMITI A5.2.3.

75 On deforestation, see Erkkilä and Siiskonen, *Forestry in Namibia*; M. Seely and A. Marsh (eds.), *Oshanas: Sustaining People, Environment, and Development in Central Ovambo* ([Windhoek], 1992).

76 For Ovamboland, see Seely and Marsh, *Oshanas* and R. J. Rodin, *The Ethnobotany of the Kwanyama Ovambos* ([Missouri]: Missouri Botanical Garden, 1985), p. 34. Cf. the seminal articles on the same fruit tree species in Zimbabwe by Wilson and Campbell. See Wilson, "Trees in Fields in Southern Zimbabwe," pp. 369–383, and B. M. Campbell, "The Use of Wild Fruits in Zimbabwe," *Economic Botany*, 41 (1987), pp. 375–385. Although they stressed that fruit trees were "wild" trees and that the fruit trees were spared when the original vegetation cover was cleared for settlement, Wilson (p. 373) and Campbell (p. 383) noted, however, that some of the fruit trees had been planted. On the basis of his research in Nigeria, Okafor comes to the same conclusion. See J. C. Okafor, "Case Study," pp. 7–17.

PHOTO 1. Trees in fields: a family weeds its fields in the shadow of towering marula trees (photo by the author, 1992).

in the 1890s as they were in the 1990s "wilderness." Close associations between tree vegetation and (dense) human settlement are, of course, far from unique to twentieth-century Ovamboland. They are at the core of Leach and Fairhead's argument about village forests in West Africa. In East Africa, McCann notes that "[p]erhaps, like modern Ethiopia, trees and permanent vegetation were a feature of the towns... marks of concentrations of humans and not nature." In the tropical regions of the Americas, various fruit and other trees are associated with abandoned settlement sites, including the Ramón tree (*Brosimum alicastrum*) and the Brazil nut.[77] Equally, durian fruit trees (*Durio zibethanus*) mark old settlement sites in West Kalimantan.[78]

But the reforestation history does not negate or redeem the history of deforestation that preceded and accompanied it; nor do the deforestation episodes invalidate the reforestation episodes: twentieth-century Ovamboland's inhabitants were neither evil Saramon-esque forest destroyers,

77 For Ethiopia, see McCann, *Green Land, Brown Land, Black Land*, p. 43. For the Americas, see Smith et al., *Tropical Forests and their Crops*, pp. 38–185, 240–242, 364–369 (Annatto), 384–394 (Brazil nut), 423 (Ramón tree). A similar tree in Asia is the Polynesian chestnut (Inocarpus fagifer), ibid, p. 423

78 Peluso, "Fruit Trees and Family Trees in an Anthropogenic Forest," pp. 54–102.

nor benign hobbit-like forest champions. The story of trees-in-fields in north-central Namibia suggests a much more ambiguous phenomenon: its inhabitants both *deforested* and *reforested* their farm lands.[79] After clearing their lands, they actively and/or passively propagated select fruit trees that within a few decades shaded their homesteads and fields. The majestic fruit trees looked so natural that colonial and postcolonial experts presumed they were the relics of a past high forest vegetation, highlighting the importance of the notion that the environmental history of today's landscapes can easily be misread. But that does not mean that colonial observers were entirely dyslexic. The deforestation in the middle floodplain was dramatic and destructive. A blanket reading of the outcome (trees-and-people) against the grain (i.e., considering the trees as the products of people's benefice) as per an inclinist perspective, rather than considering the trees as the relic of a people-free past, is enlightening conceptually, but it does not prove the point. At best, it offers a linear causality and it erases select ambiguities and contradictions. For one, whether deforestation or reforestation is privileged, choosing either as the outcome negates the other. Additionally, outcome and baseline may be blurred: descriptions of *early* and *late* twentieth-century north-central Namibia's landscapes are eerily similar, deceptively suggesting circular change, the lack of any fundamental historical change, or a benign local environmental stewardship. The contrary, however, was true: the region experienced dramatic environmental transformations in less than a century with waves of simultaneous and successive "deforestation" and "reforestation."

On-farm tree removal and propagation were first and foremost processes that were located at the intra-household, farm, field, and tree levels. It is important to highlight the need to differentiate environmental change at the microlevel. Different individuals and households varied in the extent to which they influenced environmental change by choosing which trees to cut down and which trees to nurture and how to nurture them. Only the microlevel of analysis illuminates the motivations behind human agency. For example, trees were overwhelmingly cut and propagated by individuals or by groups of individuals – processes that occurred tree by tree. Moreover, this chapter highlights that the process of on-farm tree propagation in north-central Namibia had a very strong gender dimension. Women disinvested from crop cultivation as it became increasingly male "owned" (but commensurately dependent on female

[79] For more details on the deforestation of farm and village sites, see Chapter 3.

labor) and they reinvested in propagating and protecting especially fruit trees, which were not only a relatively uncontested source of household food but also a potential source of extra household income and a form of social security.

This chapter also demonstrates that only a focus on who and what drove the cutting and propagating of trees and how and why it was executed reveals the intricacies of the processes of environmental change, for the worse *and* for the better. An explicit focus on the plural processes of environmental change reveals humans as "architects of Nature," working *with* Nature, not *against* Nature, or *in harmony* with Nature because they created, configured, maintained, and remade environmental infrastructure in interaction with other local, regional, and global actors, factors, and processes. They cut trees and actively propagated trees through seedlings and cuttings and they actively nurtured and protected young fruit trees they believed God or Nature had provided, as well as the trees they planted themselves. Rather than casting the environmental history of north-central Namibia in terms of deforestation or reforestation, the history of on-farm woody vegetation changes in the region can perhaps be more fruitfully understood as multiple processes of (re-)creating and (re-)imagining environmental infrastructure.

3

Dark Earths

Field and Farm Environmental Infrastructure[1]

Contrary to the conventional wisdom that old farms are marked by depleted soils and therefore should be shunned, Brazilian farmers in the Amazon, Indonesian farmers in Kalimantan, Dutch farmers in Frisia, East African farmers in the Great Lakes region, and Namibian farmers in the Ovambo floodplain highly prized old farm sites.[2] Brazilian farmers preferred to grow their (tree)crops on the sites of abandoned settlements and fields, just as Kalimantan swidden cultivators sought out secondary forest plots that had been cultivated before. Dutch farmers paid a premium price to ship the earth from old settlement mounds to their fields as fertilizer, East African farmers chose old homestead sites for intensive

[1] I am grateful to Susanna Hecht and James Fairhead for their encouragement in pursuing the dark earth argument in the context of southern Africa. I am also indebted to Susanna Hecht and Kathleen Morrison for giving me the opportunity to present some of my ideas at the Social Life of Forests Conference they organized at the University of Chicago on May 30–31, 2008.

[2] On the conventional wisdom, see, e.g., Schroth and Sinclair, "Impacts of Trees on the Fertility of Agricultural Soils,"in G. Schroth and F.L. Sinclair (eds.), *Trees, Crops and Soil Fertility: Concepts and Research Methods* (Wallingford, UK: Cabi Publishing, 2003), pp. 6–7, and Longman and Jeník, *Tropical Forest and Its Environment*, pp. 20–21, 24, 27–29. See also Carter and Dale, *Topsoil and Civilization*, who argue that with very few exceptions humans historically depleted and eroded soils, causing the fall of civilizations throughout history. Limbrey argues that natural and human factors cannot be easily separated in soil formation processes; see S. Limbrey, *Soil Science and Archaeology* (London: Academic Press, 1975), pp. 205–207. For critiques of the conventional declinist soil erosion narrative in Africa, see C. Reij, I. Scoones, and C. Toulmin (eds.), *Sustaining the Soil: Indigenous Soil and Water Conservation in Africa* (London: Earthscan, 1996); Scoones (ed.), *Dynamics & Diversity*; and Showers, *Imperial Gullies*.

farming; and Namibian farmers offered a hefty fee to obtain the lifetime occupancy rights to old farms when they became vacant.[3]

As such, valuing old agricultural lands is not entirely exceptional. The soils of older irrigated rice fields in Southeast Asia, e.g., are frequently more fertile than new fields, which are, in turn, more fertile than the very same soils before they were cultivated or the surrounding uncultivated soils. Continuous cultivation under submerged conditions causes changes that warrant the existence of a separate and anthropogenic "paddy soil" type, irrespective of the parent material from which the soils are derived.[4] The Amazonian, West Kalimantan, Frisian, Great Lakes, and Ovambo cases, however, suggest that the creation of valuable fertile soils on the sites of old settlements, farms, and fields is not limited to irrigated rice cultivation, but also occurred under conditions

[3] On Brazilian farmers, see W. I. Woods, "Development of Anthrosol Research," and L. German, "Ethnoscientific Understandings of Amazonian Dark Earths," in J. Lehmann, D. C. Kern, B. Glaser, and W. I. Woods (eds.), *Amazonian Dark Earths: Origin, Properties, Management* (Dordrecht: Kluwer, 2003), p. 3 and pp. 184–188, respectively. On Indonesian farmers, see M. R. Dove, *Swidden Agriculture in Indonesia: The Subsistence Strategies of the Kalimantan Kantu'* (Berlin: Mouton, 1985) and R. K. Puri, "Postabandonment Ecology of Penan Forest Camps: Anthropological and Ethnobotanical Approaches to the History of Rain-Forested Valley in East Kalimanatan," M. R. Dove, P. E. Sajise, and A. A. Doolittle (eds.), *Conserving Nature in Culture: Case Studies from Southeast Asia* (New Haven, CT: Yale University Southeast Asia Studies, 2005), pp. 25–82. On Dutch farmers, see M. Bierma, A. T. Glason, E. Kramer, and G. J. Langen (eds.), *Terpen en Wierden in het Fries-Gronings Kustgebied* (Groningen: Wolters-Noordhoff/Forsten, 1988); A. Carmiggelt, *Begraven op de Terpen in Ferwerderadiel: Het Dodenbestel in Middeleeuws Fryslan, ca. 400–1200* (Abcoude: Uniepers/Rijksdienst voor het Ouheidkunding Bodemonderzoek, 2000); O. S. Knottnerus, *Natte Voeten, Vette Klei: Oostelijk Fivelingo en het Water* (Bedum: Profiel, 2008); and E. Knol, "Wonen op Wierden: Aanpassen aan een Vruchtbaar Land nabij Zee," *Canon van Groningen: 40 IJkpunten uit de Groninger Geschiedenis* (Groningen: Huis van Groninger Cultuur/Groninger Archieven, 2008), pp. 6–15. Schoenbrun notes that old abandoned homestead sites were favored for intensive (banana) farming; see Schoenbrun, *A Green Place, A Good Place*, p. 82. Scoones also noted that old settlement sites (gardens and enclosures) were highly valued; see Scoones (ed.), *Dynamics & Diversity*, p. 30. In Central Kenya, old homestead soils were considered to be "exceptionally fertile." See A. Fiona and D. MacKenzie, *Land, Ecology, and Resistance in Kenya, 1880–1952* (Edinburgh: Edinburgh University Press, 1998), p. 50.

[4] Bray, *The Rice Economies*, pp. 28–29. See also N. C. Brady, "Soil Factors That Influence Rice Production"; Zi-tong, "On the Genetic Classification of Paddy Soils in China"; and F. R. Moormann, "The Classification of 'Paddy Soils' as related to Soil Taxonomy," in Institute of Soil Science, Academica Sinica, *Proceedings of Symposium on Paddy Soil* (Beijing: Science Press, 1981), pp. 1–19, 129–138, and 139–150, respectively. Lindert argues that in twentieth-century China and Indonesia, older fields were more fertile than newer fields and that sustained cultivation improved soil quality. See Lindert, *Shifting Ground*, pp. 187–199, 236, 239–240.

of dryland agriculture (i.e., rain-fed cultivation in both wet and semiarid environments).

In addition, it was not merely the fertile soils that made old farms and fields valuable but also the farm and field infrastructure, which critically shaped soil processes. Irrigated rice cultivation depends on an elaborate infrastructure of dams and ditches, the construction of which "requires years of hard work.... It is not surprising, then, that rice farmers often prefer to work existing fields more intensively rather than opening up of new fields, which, at least for the first few years, will produce less than long-established ones."[5] In fact, the construction and maintenance of the irrigation infrastructure created the very conditions required for the formation of fertile soils. Comparable dynamics operate in dryland cultivation. West Kalimantan dryland swiddeners and Malagasy slash-and-burn cultivators preferred the sites of former farms and fields because the secondary forest that covered them was easier to clear than primary forest, underlining the fact that land clearing requires substantial investment.[6] Moreover, the secondary forest constituted valuable environmental infrastructure that greatly facilitated cultivation.

This does not mean that old farms always and per definition were highly valued: when and where farmers did not invest in their lands, any environmental infrastructure deteriorated as homesteads and fences collapsed, wells silted up, soil fertility declined, and animals, weeds, and bush invaded the fields. But, in general, the sites of old dryland farms and fields consisted of lands that either had been cleared already or that were easier to clear because the vegetation had been altered by the previous occupants. Old functional dryland farms also often contained fruit trees or orchards, water supplies, living quarters, storage facilities, and fences to limit predation and keep out pests. Such sites sometimes also boasted conventional infrastructure (i.e., roads and trails, bridges, schools, health posts, and (rain)shrines and other places of worship). More often than not, environmental infrastructure, and even technical infrastructure, was mostly composed of organic materials drawn from the surrounding areas, resulting in the accumulation of organic matter on the farms and in the fields.

The accumulation of soil and vegetative matter from the surrounding area to protect Roman and Medieval era farmers, their animals, crops,

5 Bray, *The Rice Economies*, p. 29.

6 On West Kalimantan, see Dove, *Swidden Agriculture in Indonesia*. On Malagasy slash-and-burn, see Kull, *Isle of Fire*, pp. 148–149.

and other possessions from flooding by the sea, e.g., created the Frisian settlement mounds. The Frisian mounds were both the outcome of a process of and a means to accomplish environing – the mounds spurred further investments precisely because they offered security from the sea. The construction of dykes and the drainage of the coastal lands initiated in Medieval times, however, made the mounds superfluous and farmers abandoned them, only to have Dutch farmers carry the fertile mound earth to their fields 500 years later. In the Amazon, soil mining has also become a problem to the survival of pre-Columbian sites: its dark earth is sold in towns as gardening soil.[7]

In terms of their high organic matter content and the abundance of human settlement debris, the Frisian mounds' soils are very similar to the dark earth sites of the Amazon. Frisian farmers sought out elevated sites in the coastal floodplain. Patches of black earth in the Amazon are similarly located on higher ground, including bluffs in river floodplains, and they are surrounded by patches of brown soils. The Amazonian patches of black earth are littered with pottery fragments and other debris associated with human settlement and the lighter colored (brown) earths are interpreted as relic crop fields, highlighting the anthropogenic nature of these soils. High soil carbon contents are, to a considerable degree, responsible for the darker color of the soil patches as compared to the surrounding soils. Black and brown earths were created before the European military and biological conquests of the Amazon; the resulting demographic collapse caused the abandonment of the dark earths.[8] Central American Maya farmers similarly created fertile soils in and around their towns over a period of 3,000 years.[9] The phenomenon is also not unknown in India,

7 Bierma et al., *Terpen en Wierden;* Carmiggelt, *Begraven op de Terpen in Ferwerderadie*; Knottnerus, *Natte Voeten, Vette Klei*; Knol, "Wonen op Wierden." On soil mining in Amazonia, see E. G. Neves, J. B. Petersen, R. N. Bartone, and C. A. da Silva, "Historical and Socio-Cultural Origins of Amazonian Dark Earths," in Lehmann et al., *Amazonian Dark Earths*, p. 46.

8 T. P. Myers, W. M. Denevan, A. Winklerprins, and A. Porro, "Historical Perspectives on Amazonian Dark Earths," and Neves et al., "Historical and Socio-Cultural Origins of Amazonian Dark Earths," in Lehmann et al., *Amazonian Dark Earths*, pp. 15–28 and pp. 29–50, respectively. On the Amazonian back earths as an inspiration for carbon sequestration to benefit organic farming, see J. Bruges, *The Biochar Debate: Charcoal's Potential to Reverse Climate Change and Build Soil Fertility* (White River Junction, VT: Chelsea Green Publishing, 2009).

9 T. Beach, S. Luzadder-Beach, and N. Dunning, "A Soils History of Mesoamerica and the Caribbean Islands," in J. R. McNeill and V. Winiwarter (eds.), *Soils and Societies: Perspectives from Environmental History* (Isle of Harris, UK: The White Horse Press, 2006), p. 61.

the fertile sandy loam soils of Gujarat's Kheda district are considered to be human-made: "the outcome of several hundred years of meticulous cultivation practices: nature painstakingly reworked."[10] Mounds marked the sites of villages in the floodplains of the Orissa and Bengal deltas of India and Bangladesh, respectively. Human-made mounds dot the Mesopotamian plains of northern Syria where associated soils are distinguished by higher organic matter levels.[11] Scoones draws attention to dark-colored garden plots in Ethiopia and what he calls "hot spots" of soil fertility, including the sites of old cattle pens, gardens, and settlements. Dung accumulations mark old settlement sites in southwestern Botswana, some of which may be more than 1,000 years old.[12] An authoritative in-depth study on savanna ecosystems in South Africa's Nylsvley found that one-sixth of the study area was covered with more fertile topsoil despite infertile sandy subsoil derived from underlying sandstone parent substrata. The sites with the enriched topsoil were associated with old village sites located on the slopes of low sandstone hills.[13]

Although there is little doubt that the Frisian mounds and the Amazonian dark earths were the product of human activities, relatively little is known about the details of the processes that created them. The Frisian and other dark earth formations associated with Roman and Medieval era sites in northern Europe and Amazonian dark earths were formed hundreds of years ago and the archaeological record is limited. The formation of dark earths in the Ovambo floodplain and the surrounding areas, however, continued until very recently, and the wealth of available archival and oral history documentation allows for an analysis and description of the processes involved. The rich historical record on the Ovambo floodplain not only provides detailed insights into the agency and mechanics of dark earth formation but also highlights why they

10 V. Gidwani, "Labored Landscapes: Agro-ecological Change in Central Gujarat, India," in A. Agrawal and K. Sivaramakrishnan (eds.), *Agrarian Environments*, pp. 216–247, especially 237.

11 On the Bengal region, see D'Souza, *Drowned and Dammed*, pp. 97–99. On northern Syria, see W. Parry, "Satellites Spy Thousands of Ancient Human Settlements," Livescience.com, 3/19/2012, http://news.yahoo.com/satellites-spy-thousands-ancient-human-settlements-191006436.html.

12 Scoones (ed.), *Dynamics & Diversity*, p. 27. On Botswana, see M. Hall, *Farmers, Kings, and Traders: The People of Southern Africa, 200–1860* (Chicago: The University of Chicago Press, 1990 [1987]), pp. 86–87.

13 Scholes and Walker, *An African Savanna*, pp. 38–46. The Amazonian dark earth literature has inspired research into the origins of dark colored Roman-Medieval urban deposits across northern Europe. See P. Goldberg and R. I. MacPhail, *Practical and Theoretical Geoarchaeology* (Malden, MA: Blackwell, 2006).

constitute environmental infrastructure and how they are the product of environing.

In brief, the case of the Ovambo floodplain offers new insights into the why, where, and by whom of the processes of dark earth formation and is suggestive of the history of dark earth formation in the Amazon and Frisia as well as elsewhere in the world. It also highlights the extent to which the dark earths concerned are not an artifice exclusively created by and for humans.[14] These dark earths constitute *environmental* infrastructure because their creation is only in part shaped and dominated by humans: dark earth formation is dependent on in situ chemical and other processes as well as on significant off-site nutrient imports. And this draws attention to a "darker" side of the processes of "dark earthing" that is sometimes lost in inclinist narratives. Locales with "anthropogenic" dark earths gained fertility at the cost of draining nutrients from other sites where the effect was environmental impoverishment, highlighting that multiple and contradictory processes were at work. A focus on the history of "anthropogenic" dark earths as *environing*, therefore, does not only problematize the view that (indigenous or premodern) soil use per definition results in environmental degradation (as per a modernization or declinist perspective); it equally highlights that indigenous soil use does not "naturally" constitute environmental improvement or sustainability (as per an inclinist model).

Environmental Infrastructure and Dark Earth Spaces

Similar to their occurrence in Frisia and Amazonia, "anthropogenic" dark earths in the Ovambo floodplain are limited to highly specific locations in the landscape: on the higher ground in the low sandy ridges that are sometimes referred to as "relic dunes." The "natural" dark earths of the lower-lying seasonal river beds and pans were enriched by silt deposits and accumulations of organic matter but were flood-prone and often too saline. Just as in Frisia and Amazonia, the threat of flooding in the Ovambo floodplain limited settlement and cultivation to patches

[14] McCann referred to soils as "actors" because soils move through erosive processes; see McCann, *Green Land, Brown Land, Black Land*, p. 12. A new paradigm in soil sciences calls for a reevaluation of the importance of organic as opposed to inorganic fertilizers in sustaining soil fertility in Africa. See B. Vanlauwe, J. Wendt, and J. Diels, "Combined Application of Organic Matter and Fertilizer," in W. A. Dick et al., (eds.), *Sustaining Soil Fertility in West Africa* (Madison, WI: Soil Science Society of America and American Society of Agronomy, 2001), pp. 247–279.

of land that were above the flood line. These are, however, not necessarily entirely "natural" niches. In Amazonia, many dark earth sites are on bluffs; human activity increased the height of the bluffs. Frisian farmers equally took advantage of natural high ground, but they also intentionally built up the sites they chose for settlement by adding clay and clods to create mounds. Building up sandy soils with peat, moss, heather, or grass sods mixed with manure was common in many areas across Medieval Western Europe from Ireland to Germany, creating the so-called plaggen soils that are noticeably raised over the surrounding landscape.[15] Similarly, Ovambo floodplain farmers increased the height of the dunes when they formed dark earth soils on their farms and fields. Floodplain farmers in the Chinese Yangtze delta, the Egyptian Nile valley, northern Italy, Pakistan's Punjab, and the Orissa delta of India, took advantage of the annual floods to build up their fields' soils: the waters deposit fertile silt eroded upstream.[16]

Fields and farms are typically the rural space where the most radical environmental changes occur. The land is cleared with the axe, machete, or firebrand, and wood and other materials harvested on and around the farm plot serve to construct huts, storage facilities, fences, palisades, and other farm infrastructure. The "deforestation" footprint that results from clearing and constructing farm and field infrastructure often extends beyond the actual farm plot. Palisades, fencing, herding, hunting crop pests, and the use of fire and broom also protect the farm plot from intensive use by most animals, from the very large to the very small, creating a space where animal impact is heavily managed through spatial and temporal patterns of exclusion. Paradoxically, the farm plot is at once the site with the most severe deforestation as well as the locus of the most

[15] On plaggen soils, see P. Hoppenbrouwers, "Agricultural Production and Technology in the Netherlands, c. 1000–1500," and M. Widgren, "Fields and Field Systems in Scandinavia during the Middle Ages," in G. Astill and J. Langdon (eds.), *Medieval Farming and Technology: The Impact of Agricultural Change in Northwest Europe* (Leiden: Brill, 1997), respectively pp. 89–114 and 173–192. See also Limbrey, *Soil Science and Archaeology*, pp. 335–336, and Smout, *Exploring Environmental History*, p. 139. In fifteenth-seventeenth-century coastal Holland, entrepreneurs mined dune sand to raise soils in the swampy hinterlands for urban construction and crop fields. See P. J. E. M van Dam, "Rabbits Swimming Across Borders: Micro-Environmental Infrastructures and Macro-Environmental Change in Early Modern Holland," in Scott G. Bruce (ed.), *Ecologies and Economies in Medieval and Early Modern Europe* (Leiden: Brill, 2010), pp. 63–91. For indigenous Maori "plaggen soils," see B. C. McFadgen, "Maori Plaggen Soils in New Zealand, Their Origin and Properties," *Journal of the Royal Society of New Zealand*, 10 (1980), pp. 3–19.

[16] D'Souza, *Drowned and Dammed*, pp. 43–44.

intense reforestation.[17] Moreover, actual or former farm plots frequently are marked by high levels of biodiversity. House gardens with fruit trees, orchards, and agroforestry occur on farms and in fields and are ubiquitous worldwide. The on-farm trees constitute important environmental infrastructure in and by themselves, supplying fruit, fodder, construction materials, and tools. The trees also provide a range of services, including shading humans, animals, and plants, reducing temperature, marking property and boundaries, and hosting bees and spirits.

Preparing and using the land for settlement and cultivation also critically shapes soils. Human activities that consciously or unconsciously alter the physical structure or chemical composition of soils lead to soil losses and/or gains. For example, clearing vegetation exposes the soil to higher levels of solar radiation, raising soil temperatures and making the topsoil more vulnerable to water and wind erosion. Soils exposed to erosion and subjected to continuous cropping may impoverish to a degree that they can sustain little or no vegetation, leading to desertification.

But settlement and cultivation sites may also improve soils because such locations often accumulate organic and inorganic matter from the surrounding areas. This may entail both mechanical and chemical processes. People may deliberately accumulate soil in particular sites for religious purposes (e.g., temple mounds), for defense against attack or flooding, or to create deeper and more fertile soils to grow crops or trees. Sustained use of settlement or crop locations may result in mounds because of the accumulation of household and other debris over time. Upland and bottomland use enhances horizontal and vertical mineral and nutrient transportation through natural and human-made drainage channels, including irrigation canals, roads, and paths into (e.g.) valleys, ditches, reservoirs, ponds, and paddies. Accumulating soils and adding inorganic and organic matter, as well as cultivating, draining, or irrigating soils also alter their physical and chemical composition. Flooding rice fields, e.g., inhibits weed growth, irrigates rice plants, and enhances mineral and nutrient recycling, fertilizing the soil.[18]

The Amazonian dark earths are high in organic matter and associated with the remains of human settlement. They are highly prized by both indigenous and non-indigenous Amazonians and they are the sites of choice to establish fields or orchards. The darker black earths, high

[17] See Chapter 2.

[18] See Bray, *The Rice Economies*. On religious and burial mounds, see, e.g., G. R. Milner, *The Moundbuilders: Ancient Peoples of Eastern North America* (London: Thames and Hudson, 2004).

in carbon, are thought to be past settlement sites marked by the debris from habitations, fires and hearths, and middens. Lighter brown earths resulting from crop residue and nutrient imports mark old fields. The Amazonian dark earths have pre-Columbian origins and the process of dark earth formation rarely seems to have continued beyond the European military and biological invasion of the region and the subsequent Amerindian demographic collapse.[19]

As in Amazonia, the formation of dark earths in north-central Namibia is concentrated in villages on the sites of farms and fields. In fact, the very layout of farms and fields, demarcated and segregated from the larger environment by fences and palisades, shapes *where* dark earths are formed. Farmers' use and management determine *how* and at *what rate* dark earths are formed. Dark earths form on farms and fields because the farm plots constitute mineral and nutrient sinks that are privileged over the surrounding environment: farmers' land use practices transport nutrients and minerals to the farm plots. Down-slope fields receive eroded soil from higher up and downstream fields may receive fertile silt carried from upstream. Construction materials for huts, stores, tools, palisades, and fences are partly sourced beyond the farm plot, and the decaying construction materials themselves contribute substantially to the formation of dark earths. Wooden and clay farm constructions are replaced and renewed every couple of years and the location of homesteads (huts, stores, walls, and palisades), and livestock kraals often are rotated through the farm plots, spreading the dark earths. Animal and human excrement and urine are also deposited on-farm. In north-central Namibia, goats are penned on-farm each night throughout the year; the same is true for cattle during the rainy season with fences or herders preventing their access to the actual fields. The soil also is mechanically manipulated to create raised fields to enhance soil structure and composition and to minimize the impact of water and wind erosion while optimizing crop productivity.

Huts and Fences as Environmental Infrastructure

An outer fence that enclosed the fields, fruit trees, a fallow, and the farm homestead with its huts and palisade often marked individual farms in Ovamboland. The wooden infrastructure of palisades, roofs, walls, and

[19] On dark earths in the Amazon, see Lehmann et al., *Amazonian Dark Earths* and B. Glaser and W. I. Woods (eds.), *Amazonian Dark Earths: Explorations in Space and Time* (Berlin: Springer, 2004).

fences protected families, domestic animals, vegetation and soils, and food and seed stores from human and animal predation, vermin and weeds, and the elements. The palisades and fences not only safeguarded precious resources but also encouraged further and cumulative investments by enhancing security. Petra Van Dam adopts the term "microenvironmental infrastructure" to designate similar elaborate fences to protect villages, farms, horticultural gardens, and crop fields from rabbits in the dune landscapes of western Holland in the fifteenth through seventeenth centuries. Rabbits were the property of the counts of Holland and could not be killed even if they invaded private gardens and fields and damaged crops. Keeping out the rabbits forced villagers and farmers to invest into erecting and maintaining wooden fences consisting of expensive imported wooden panels that went 30 cm (one foot) down in the soil and 1.5 meters (five feet) up.[20]

In the northern floodplain Oukwanyama kingdom (in modern Angola) in the 1860s, the size of a farm plot cleared in an area with few or no neighboring farms was limited only by the labor capacity of the household, as is captured in a recorded saying: "the hoe determines the borders of the field."[21] A fence of thorny branches surrounded the farm of a homestead and its corresponding fields and livestock enclosures. In the early 1880s, the fenced farm of one Ombadja king was so expansive that his European visitors compared it to a plantation.[22] The homesteads commonly were located in the middle of the crop fields, except in southern floodplain Ondonga, where the homesteads were constructed on the edge of the fields.[23] In some areas, farmers situated their farms and fields around pans, depressions that held water during the rainy season.[24] In 1917, in the southern floodplain villages, fields of neighboring farms were located closely together, separated only by narrow tracks. Ondangwa in

20 Wood was a scarce commodity in Holland. The underground wood consisted of recycled panels from old ships, but the above-ground wooden panels had to be imported. See van Dam, "Rabbits Swimming across Borders," pp. 63–91.

21 A. Wülfhorst, *Shiwesa, Ein Simeon aus den Ovambochristen* (Barmen: Verlag des Missionshauses, 1912), pp. 7–8, and A. Wülfhorst, *Moses, Eine Erstlingsfrucht aus den Ovambo* (Wuppertal-Barmen: Missionshaus Verlag, n.d.), pp. 14–15.

22 NAN, NAO 104, C. L. Anderson to Hahn, Jordan diary.

23 AVEM, RMG 2599 C/i 19, Bernsmann, Omburo, January 6, 1892, and RMG 2518 [?] C/h 52, Speiker, Visitationsbericht, Namakunde, July 13–18, 1906; Möller, *Journey*, pp. 110–111, 126; NAN, NAO 104 Anderson to Hahn, Jordan diary and A233, J. Chapman, 1903–1916, 1876[?] [pp. 61–62], and A450 Vol. 6, Intelligence Report no. 10, Namakunde 3/1/1917.

24 Lau, *Carl Hugo Hahn Tagebücher*, vol. iv, July 22–23, 1857, pp. 1040–1045.

the Ondonga kingdom was "surrounded by fields and kraals," and, during a 1917 military campaign to crush the resistance by the Oukwanyama King Mandume, the South African forces were intentionally stationed in the uninhabited flats southwest of the Ondonga kingdom's villages to prevent crop destruction and infringement on "native rights."[25]

The construction of new farms' fences in the middle floodplain, in the western half of the southern floodplain and east of the floodplain by the various waves of refugees and migrants, consumed a considerable amount of bush.[26] Fencing was less common in the southern floodplain as is underscored by frequent complaints that livestock had damaged crops in colonial-era Ondonga – and to a lesser extent in Ongandjera further west. Under such circumstances, farmers had to rely on herdsmen – often young boys – to keep livestock from entering their crop fields during the rainy season. A mid-1920s manuscript that was probably based on information obtained from a missionary in Ondonga emphasized: "[i]n the rain season the cattle must be strictly shepered [sic: shepherded], for the cornfields are without any fence."[27] In 1993, however, only one of every six sampled households in north-central Namibia lacked any form of farm fencing.[28]

Late nineteenth-century visitors were impressed not only by the layout and the size of the floodplain villages' crop fields but also by their

25 NAN, RCO 10, RCO and Hahn, Preliminary memo re[garding] Ovamboland and Chief Mandume, [1916] and RCO to staff officer Union forces SWA, Ondonga, December 3, 1916.

26 NAN, A450, 7, Annual Report Ovamboland 1937; NAO 36, Annual Health Report 1937; NAO 46, Rex vs. Ninda, Statement ANC, March 9, 1939, statement Kenatuka Keivanga, March 9, 1939, Rex vs. Chiwetha Hiyanga, Chief Warden Windhoek Gaol to NCO, Windhoek, November 23, 1942; NAO 45, statement Mattheus Angula, [Ondangwa], September 14, 1939; NAO 17, NCO to Secretary SWA, Ondangwa, August 26, 1940; A450, vol. 23, D6, Land Tenure; NAO 65, Annual Health Report Ovamboland 1952; OVA 49, Meeting Subcommittee Townplanning, September 2, 1970. For an example of a fenced farm in Onkolonkathi, see NAO 46, statement Shiwetha Amawha, Ondangwa, February 3, 1941. Thornbush was common in the northwest, see, e.g., NAO 18, Monthly Report Ovamboland, September 1925 and A450, vol. 23, D6, Land Tenure [Eunda].

27 NAN, A450, vol. 9, Anonymous, "Ovamboland Customs (1926)"; RCO 10, RCO to MacKenzie, Ondonga, February 9, 1917; NAO 18, Monthly Report Ovamboland, August 1926; NAO 9, blue note with heading "Chief Martin," n.d. but inserted between letters dated April 6 and August 26, 1929; NAO 9, Bjorklund to NCO, Onajena, June 10, 1934; NAO 100, NCO to Chief Kambonde, Ondangwa, December 8, 1952, and Chief Kambonde to NCO, Okaroko, December 15, 1952; NAO 91, Kambonde to NCO, Okaroko, January 14, 1953; NAO 46, statement Shikua Iyambo, appendix NCO to Attorney-General, Ondangwa, September 9, 1941.

28 OMITI 4.3.6.

productivity: grain surpluses were stored in huge baskets and food was readily offered for sale, and in abundant quantities.[29] During the late pre-colonial era, millet and sorghum were the floodplain staples, with beans, pumpkins, and bambara nuts as secondary crops. During the dry season, stored grains were often the only food available.[30]

The 1915 Portuguese invasion of the northern floodplain caused a massive refugee exodus into the southern and middle floodplain at the time that South African forces entered the area, having defeated the German colonial army and conquered the German colony of South West Africa. Ovamboland's long-serving Native Commissioner C. H. L. Hahn first arrived in the Ovambo floodplain in 1916 as a junior military officer in the South African army. The delimitation of the Angolan-Namibian boundary in 1927 spurred a second wave of refugee movements from Angola into colonial Namibia's middle floodplain. The effect on the environment was dramatic: "[n]ew kraals [homesteads] are seen everywhere and many natives are busy felling trees and clearing places for further settlement."[31] Native Commissioner Hahn commented on the expansion of settlement in the middle floodplain that led to the creation of the new colonial Oukwanyama district in the former wilderness, an area he had known before the refugee influx: "many patches have been denuded of trees and cultivated." Hahn's subaltern in the Oukwanyama district wrote that "the destruction in ukuanyama [Oukwanyama] of forest trees has been carried on at an alarming rate during the last ten years."[32]

Hahn favored the settlement of what he considered the wilderness areas of north-central Namibia. But he wanted to stamp out "shifting cultivation" as wasteful. He, therefore, engaged the chiefs and headmen to urge settlers who cleared new land not to burn the large trees on

[29] Lau, *Carl Hugo Hahn Tagebücher*, July 22–23, 27, 1857; AVEM, RMG 2599 C/i 19, Beersmann, Omburo, January 6, 1892; NAN, A233, J. Chapman, 1903–1916, 1876[?] [pp. 61–62] and NAO 104 Anderson to Hahn, Jordan diary; Möller, *Journey*, p. 110.

[30] Lau, *Carl Hugo Hahn Tagebücher*, July 22–23 and 27, 1857; NAN, A233, J. Chapman, 1903–1916, 1876[?] [pp. 61–62] and NAO 104 Anderson to Hahn, Jordan diary; D.M. de Lima, *A Campanha dos Cuamatos Contado por um Soldado Expedicionario* (Lisbon: Livraria Ferreira, 1908), pp. 155, 159; AVEM, RMG 2518 [?] C/h 52, Speiker, Visitationsbericht, Namakunde July 13–18, 1906, and RMG 2599 CNDIH, Avulsos, Caixa 3703, Processo Missão de Estudos no Sul de Angola, 1914–1915, Relatorio do Mes de Outubro [1914].

[31] NAN, NAO 18, Monthly Reports Ovamboland, June–July 1927. See also KAB 1, W. Volkmann, "Report on the Agricultural and Political Conditions at The Angola Boundary," October 30, 1928.

[32] NAN, SWAA 3, NCO to SEC. SWA, Ondangwa, April 20, 1931.

their plots, and he recommended severe punishment for the "unnecessary destruction or mutilation of trees." Moreover, felled trees had to be used as timber and the stumps were dug out for use as firewood. In 1938, to encourage felling rather than burning timber-size trees, the administration entrusted the senior headmen of the Oukwanyama district with large axes and saws.[33]

Hahn's administration called not only for a less destructive use of the local resources but also for more efficient agricultural production beyond subsistence levels. He and his staff pressed Ovamboland's inhabitants to prepare larger fields so that food reserves could be built up to prevent a repeat of the 1929–1932 Famine of the Dams, when the colonial administration had to provide costly food aid. The 1933 annual report for Ovamboland noted with satisfaction that the recommendation had been heeded in the southwestern districts of Uukwambi and Ombalantu. Throughout the 1930s in the large Ondonga district, however, frustrated officials continued without result to pressure the king and his senior headmen to induce the district's inhabitants to cultivate larger fields. Small fields were partially a legacy of the insecurity caused by the violence and displacement that plagued the region in the 1910s and early 1920s. Displaced by the 1910s warfare in the northern floodplain, Kanime Hamyela and her family, e.g., had only a very small field in the Ondonga village where they had sought refuge. Even if they were able to do so, people simply felt too physically insecure to invest in improving or extending fields and other environmental infrastructure. In Uukwambi and Ongandjera districts, "[i]n former years the rank and file were afraid to do this [invest in larger fields] for no sooner had one cultivated a nice big patch than a stronger tribesman would, by bribing the chief or headman, deprive him of it."[34]

Information about the size of farm plots is scarce. The field of an 1880s Ombadja king measured perhaps 1,000 meters across, and may

[33] NAN, SWAA Native Affairs 456, NCO to Senior Veterinary Surgeon, Windhoek, Odongua [Ondangwa] April 24, 1932; NAO 19, Monthly Reports Ovamboland, November 1932 and June 1933; NAO 20, Monthly Reports Ovamboland, October–December 1940 and O/C NA Oshikango to NCO, Oshikango, June 20, 1938.

[34] NAN, NAO 19, Monthly Reports Ovamboland, November 1932 and June 1933; A450, vol. 12, SWA Commission: Minutes of Evidence (1935), vol. 9, Session at Otjikango [Oshikango], August 9, 1935, 518–520; A450, vol. 7 f. 2/18, Annual Report Ovamboland 1926. On small fields and general insecurity, see A450, vol. 9, Hahn, "Rough Notes on Tribal Customs in Ovamboland," p. 59 and vol. 10, "Political and Tribal Affairs," p. 3 [Mss]. Kanime Hamyela, interview by author, Omutwewondjaba (Namibia), June 15, 1993.

have been as large as 10 ha.[35] In the case of a late-1930s Uukwaluthi farm, the distance from the homestead to the fence around the fields was sixty to seventy meters, suggesting a farm size of roughly 0.5 ha.[36] In 1950, the Native Commissioner of Ovamboland estimated that the area's 27,606 households cultivated 1 ha each.[37] The agricultural report for 1955/1956 estimated the average farm plot at 1.7 ha.[38]

In 1964, the Oukwanyama district's estimated 83,000 inhabitants had 36,837 ha under cultivation (0.4 ha per inhabitant on average, or 1.6 ha for a family of 4), compared to 3,344–13,378 ha in 1933, when the population was 34,000 inhabitants. In 1915, when the area was still considered to be wilderness, it contained at most 600 households with 300–1,200 ha of farmlands. The land taken up by farms was but a fraction of the total surface of the district. The total surface of the Oukwanyama district in 1964 was estimated at 565,408 ha. Thus, in 1915, 0.05–0.2 percent of the total land surface consisted of farmlands. The estimated figure for 1933 is 0.6–2.4 percent, and 6.5 percent for 1964.[39]

Newly constructed farms often had small fields initially. Many of the early settlers' farm plots sloped gently toward the seasonal watercourses and the plots were located on the edge of the water course. On the uphill side, the farm plot bordered the mixed tree and bush patches on top of the ridges. Many farm plots seem to have been extended into the edges of these patches of woodland, and parts of the woodland were fenced in.[40]

In densely settled areas, the fields of different households often adjoined because the best farm plots were located on the edge of the seasonal water courses. In a late-1930s Uukwaluthi district example, two neighboring homesteads were only fifty to sixty meters apart. In an early-1940s case from Ongandjera district, neighboring homesteads were in full view of one another.[41] In general, during the late 1920s, in the

35 NAN, NAO 104 C.L. Anderson to Hahn, Jordan diary.

36 NAN, NAO 46, statement Kenatuka Keivanga, March 9, 1939, Rex vs. Chiwetha Hiyanga, Chief Warden Windhoek Gaol to NCO, Windhoek, November 23, 1942.

37 NAN, NAO 103, NCO, Ondangwa, October 21, 1950, and Census of Agriculture Ovamboland 1949/1950.

38 NAN, BAC 133, Agricultural Report Ovamboland 1955/1956.

39 NAN, AHE (BAC) 1/352, Annual Report Agriculture Oukwanyama 1964.

40 Nahango Hailonga, interview with author, Onamahoka (Namibia), February 4, 1993; NAN, NAO 36, Annual Health Report 1937; AVEM, RMG 2599 C/i 19, Bernsmann, Omburo, January 6, 1892. The woodland patches on the crest are known as *omufitu* (sing.).

41 NAN, NAO 46, statement Kenatuka Keivanga, March 9, 1939, Rex vs. Chiwetha Hiyanga, Chief Warden Windhoek Gaol to NCO, Windhoek, November 23, 1942,

northwestern districts of Ovamboland (which included Uukwaluthi and Ongandjera), the population density was much lower and a report noted "[k]raals are much more scattered and distributed over wider areas."[42]

These figures allow for an assessment of the labor and wood resources required for converting "wilderness" lands into farms and fields. In 1933, colonial Namibia's new Oukwanyama district, which had been almost entirely uninhabited in 1915, counted 18,000 homesteads. Presuming an average farm size of 0.5 to 2 ha, the refugees from the northern floodplain who settled in the new district must have cleared 9,000 to 36,000 ha of forest for their farm plots alone. Constructing the huts, palisades, and fences, however, required another 9,000–18,000 ha of mopane-type bushland to supply building materials. Thus, a total of 18,000–54,000 ha of woodlands was cleared in just more than three decades.[43] The labor required included not only cutting (and burning) the vegetation but also shaping the poles and sticks needed for the palisade and fence and the actual construction of the fences, palisades, and huts. The palisade poles were sharpened at the top and the lower part of the poles were set in a deep trench that was filled after completion.

Converting bushland into farmland demanded a major investment of time and labor and goes a long way to explain why established farms were highly valued.[44] For households short of labor, the challenge of converting a tract of bushland was daunting. Holongo Amshelelonanda from the Ondonga district stated in 1952 "[I] did not go in the unoccupied bush to make my kraal there because I am now old and find it very difficulty [*sic*] to cut the bushes and poles in order to make a new kraal for myself. My big sons who can help me are in the south [i.e., are employed outside of Ovamboland]."[45]

Moreover, converting bushland into farmland was not an event, but a process. Once established, environmental infrastructure had to be incessantly maintained, replaced, repaired, and renewed. Otherwise, huts, palisades, and fences broke, rotted, and collapsed, allowing animals and plants to invade the farm space, displacing and destroying trees and crops

and statement Shikua Iyambo, appendix to NCO to Attorney-General, Ondangwa, September 9, 1941. For Oukwanyama, see, e.g., AVEM, RMG 2599 C/i 19, Bernsmann, Omburo, January 6, 1892.

42 NAN, NAO 18, Monthly Report Ovamboland, September 1926.

43 Kreike, *Deforestation and Reforestation in Namibia*, pp. 164–165.

44 Wülfhorst, *Moses*, p. 14–15.

45 NAN, NAO 100, Chief Kambonde to NCO, August 8, 1952, and Statement Holongo Amshelelonanda at Ondangwa, July 4, 1952.

as well as the food, seed, and water stores, and thereby imperiling livelihoods and lives. Late in the dry season, usually in September, villagers cleared fields in anticipation of the rains and crop seeding. Fire was an important tool in clearing the fields of crop residue and weeds.[46] The construction and upkeep of palisades and fences as protection against animals were especially urgent in the context of growing livestock herds and the 1920s–1940s resurgence of wildlife populations – large predators and elephants in particular – in the Ovambo floodplain, coupled with disarmament that banned firearms.[47]

Fields and crops were best protected by a fence, which needed to be repaired and maintained every year. When a farm lacked a fence or the fence was broken, animals could cause severe damage. Mingana Shikongo, an elderly woman from Oukwanyama, recalled losing her entire grain crop in the mid-1940s because a neighbor's cattle repeatedly invaded her unfenced field. Kalikalelwa Oshitina Muhonghwo's 1993 plight is equally illustrative. Like Mingana Shikongo, she was elderly and alone, and unable to maintain a proper farm fence. Her fence had large holes and roaming goats from neighboring households were a constant threat to her standing crops.[48]

Kalikalelwa Oshitina Muhonghwo was also unsuccessful in keeping the bush vegetation on her farm in check. When clearing a new farm or field, people often left the tree stumps in place because removing them was arduous. To save labor, Kalikalelwa Oshitina Muhonghwo followed this custom, leaving the stumps and roots of the bush she cleared in the small plot where she planned to grow millet. But she had been unable to sow millet for the previous two years because silver cluster leaf bush that had re-sprouted from the remaining roots had re-invaded the plot. She intended to cut back the bush but added that good rains might cause the bush to re-sprout yet again, highlighting the tenacity

[46] NAN, NAO 18–19, Monthly Reports Ovamboland, September 1925, September 1926, September–October 1932, November–December 1935; NAO 60–61, Quarterly Reports Ovamboland April–September 1948, October–December 1953; Kalolina Naholo, interview by author, Ohamwaala (Namibia), January 26–27, 1993 and NAN, A450, 12, SWA Commission: Minutes of Evidence, Otjikango, August 9, 1935, Gawthorne (CEM), p. 568. See also A450, 7, Annual Report Ovamboland 1937 and NAO 64, Telegram NCO to Commandant Fourie, Ondangwa, September 25, 1953. See also the following discussion.

[47] Kreike, *Deforestation and Reforestation in Namibia*, pp. 90–107.

[48] NAN, NAO 99, NCO to ANC, Ondangwa, October 30, 1950, appendix: Statement Mingana Shikongo, October 28, 1950; Kalikalelwa Oshitina Muhonghwo in Ondaanya, interview by the author, Ondaanya (Namibia), February 2, 1993.

of the species and the labor investments required to keep a field free from bush encroachment.[49] Colonial sources also emphasize the high labor investment required to maintain farm environmental infrastructure. For example, Native Commissioner Hahn claimed in 1937 that the Christianization of north-central Namibia had left families shorthanded. Pre-Christian polygamy, he argued, encouraged large families that could maintain "grand" homesteads. In small, monogamous Christian families, however, the husband had little assistance and "[e]ventually the poles rot away . . . and he, being alone, cannot fetch new ones, so he uses corn stalks and bushes and his kraal develops into an eyesore."[50]

Because fencing and other construction materials decayed, they frequently had to be replaced. Half of the OMITI respondents emphasized that fencing materials had to be replaced every year and nearly one-third mentioned that replacement fencing was required every two to ten years. The most frequently used fencing materials were the acacia species camel thorn and sicklebush.[51]

In 1993, one of every three households surveyed mentioned that mopane palisade poles, which were the most commonly used construction material, needed to be replaced annually. But 40 percent of the respondents emphasized that mopane poles lasted for more than five years. Silver cluster leaf poles were somewhat less hardy than mopane: nearly 20 percent of the respondents posited that poles cut from the species withstood rot and decay for at least half a decade. Using silver cluster leaf as a source of poles had another advantage: once planted in the ground, the poles could develop roots and turn into a living fence.[52] That woody vegetation was the principal source of fencing and other environmental infrastructure not only shaped how the land was used spatially and temporally but also made a major contribution to soil structure and composition because decaying wood was a major component in the formation of dark earths.

49 For leaving tree stumps as a labor-saving strategy, see, e.g., Wülfhorst, *Moses*, pp. 7–22, 36–37. See also Kalikalelwa Oshitina Muhonghwo in Ondaanya, interview by the author, Ondaanya (Namibia), February 1, 1993.

50 NAN, A450, 7, Annual Report Ovamboland 1937.

51 OMITI 4.3.3, 4.3.9, 4.3.10, and 4.3.12.

52 OMITI 4.3.4. On silver cluster leaf, see interviews by the author: Kalolina Naholo, Ohamwaala (Namibia), January 26–27, 1993, Kaulikalelwa Oshitina Muhonghwo, Ondaanya (Namibia), February 1, 1993, and Philippus Haidima, Odibo (Namibia), December 9, 1992. In eastern Ovamboland, where mopane was less commonly used, Erkkilä estimated that a pole would have to be replaced after 15 years, Erkkilä, "Living on the Land," p. 91.

Making Dark Earth

Just as colonial and post-colonial officials and experts regarded north-central Namibia's fruit trees as wild and wilderness resources, with its inhabitants merely enjoying the natural bounty, they tended to see the region's agricultural infrastructure as little more than itinerant, random, and if constructed, as easily lost as replaced. To their eyes, traditional indigenous farms with their stick and grass huts, pole palisades and thorn bush fences, muddy water holes, and irregularly shaped and mixed-crop fields seemed hardly an improvement on Stone Age hunting and gathering. Native Commissioner Hahn described local agriculture as shifting cultivation (i.e., a primitive and itinerant cultivation form that was not only merely and narrowly) subsistence oriented, but also parasitic, and, moreover, utterly wasteful of natural resources). As Hahn explained in 1941:

> [t]he system of shifting cultivation is also responsible for the destruction of timber, mostly mopane which is very plentiful. Ovambos on the whole make good use of manure, but the quantity available is not sufficient and the regular clearing of new fields is unavoidable.[53]

Because Hahn was a first-hand observer of the extent to which woody vegetation had been cleared in what had been the uninhabited bushlands between the northern and the southern Ovambo floodplain societies in the late 1910s and 1920s, his use of the shifting cultivation image is not entirely surprising. After the violent Portuguese conquest of the northern floodplain, thousands of refugees fled into the middle and southern floodplain. At the same time, thousands more from the heartlands of the southern floodplain fanned out into the middle floodplain "wilderness," joining the refugees from the north in converting the bushland into landscapes of farms and fields.[54]

What Hahn and his colleagues witnessed in the 1920s and 1930s, however, was a one-time, unsynchronized process of creating new environmental infrastructure in regions that had little or no evidence of recent

[53] NAN, SWAA 3, Administration, Forestry: Indigenous Forests Ovambo A1/2 (I), NCO to CNC, Ondangwa, June 2, 1941. See also NAO 62, Agricultural Report Ovamboland 1953 and NCO to CNC, Ondangwa, March 11, 1954.

[54] NAN, NAO 10, ANC to NCO, October 31, 1940, and Hahn's handwritten notes on the letter: "Also in regard to indiscriminate burning of Mopane trees in Ukuambi and Ukuanyama lands"; SWAA 3, Administration, Forestry: Indigenous Forests Ovambo A1/2 (I), NCO to CNC, Ondangwa, June 2, 1941; Kreike, *Deforestation and Reforestation*, pp. 27–35.

prior settlement, as opposed to a structural phenomenon of shifting cultivation. Refugee colonization frontiers similar to the one established in north-central Namibia during the early 1900s occurred throughout the continent. As the violence of the slave trade and colonial conquest and pacification abated, millions of displaced Africans returned to the lands from which they had previously fled, or they settled elsewhere in "empty" lands, trying to rebuild their lives and livelihoods.[55]

Shifting cultivation has been seen not only as primitive and wasteful, but also as environmentally destructive, and a major cause of deforestation in Africa, the Americas, and Asia. Scholars of indigenous societies have argued, however, that although shifting cultivation is land-extensive, it is sustainable under low population pressure, with much less environmental impact than conventional agriculture. Some have even argued that shifting cultivation was not the dominant form of agriculture in the Americas before European conquest at all, but that it was largely a post-conquest innovation adopted as a response to the resulting violence, displacement, and depopulation.[56]

Western soil scientists conventionally consider that pristine "natural" soil contains its optimum fertility; human agriculture depletes its fertility.[57] Floodplain farmers in north-central Namibia, however, saw themselves as soil builders. This view meshes with an emerging new paradigm of soil management that is based on soil biological processes and improving soil organic matter. But the new paradigm emphasizes "natural" processes of soil organic matter buildup and largely ignores the role of humans. The Ovambo floodplain farmer's view also fits with a body of literature on Latin America and Africa that stresses the

[55] Kreike, *Re-Creating Eden*, pp. 4–10.

[56] For shifting cultivation, see Cairns (ed.), *Voices from the Forest*. For shifting cultivation as a post-conquest "innovation" in Latin America, see Denevan, *Cultivated Landscapes*, pp. 56–73. Scott argues that shifting cultivation is a state-evading strategy ("escape agriculture") practiced by refugees from state-making processes. See Scott, *The Art of Not Being Governed*, pp. 126–219.

[57] Huxley states that soil formation requires hundreds of years, and that soil losses are "effectively irreplaceable in our lifetime." See Huxley, *Tropical Agroforestry*, pp. 51, 132, 261, 265, 276–277, 280, 283. See also Longman and Jeník, *Tropical Forest and its Environment*, pp. 20–21, 24, 27–29. Most chapters in Nair present soil as natural givens; see, e.g., P. Poscher, "An Evaluation of the *Acacia albida* based Agroforestry Practices in the Hararghe Highlands of Eastern Ethiopa" and J. C. Okafor and E. C. M. Fernandes, "The Compound Farms of South-Eastern Nigeria: A Predominant Agroforestry Homegarden System with Crops and Small Livestock," in P. K. R. Nair (ed.), *Agroforestry Systems in the Tropics* (Dordrecht: Kluwer, 1989), pp. 385–400 and pp. 411–426, respectively.

anthropogenic origins of certain soils, e.g., the innovative work on the Amazon's dark earths. Anthropogenic soils may be more fertile than "natural" soils in any given environment because of their history of accumulating organic matter as a result of intentional and unintentional human action, including the application of manure and household debris (middens, decaying wood). As did Ovamboland's farmers, Mende farmers in Sierra Leone, e.g., preferred locations that had been inhabited in the past as sites for (tree)crops because they considered such areas to be especially fertile.[58] Ovamboland's farmers constructed raised beds or mounds, applied manure and other organic matter, and employed a homestead fallow system that created dark earths and sustained continuous cropping.

North-central Namibia's farmers acknowledged their expertise as architects of the soil. After having been expelled from his farm, Holongo Amshelelonanda lamented "[m]y late land is a very good land. I am the person who made in [*sic*: it?] good and fertile by manuring it."[59] In the early 1900s, a German missionary emphasized that one of the first Christian converts had made a major sacrifice by abandoning his old established farm to move closer to the mission because the farm had been very fertile. After moving to the mission, he had to clear a new plot in the

[58] A. Young, *Agroforestry for Soil Management* (Wallingford, UK: CABI, 1997 [1989]), pp. 23–25, 29, 99–110. Compare one of the published contributions to a major 2000 soil science conference that highlighted the human role in soil biological processes; see F. Ishida, G. Tian, and T. Wakatsuki, "Indigenous Knowledge and Soil Management," in Warren A. Dick et al., (eds.), *Sustaining Soil Fertility in West Africa* (Madison, WI: Soil Science Society of America and American Society of Agronomy, 2001), pp. 91–109. On anthropogenic soils in Africa, see, e.g., P. Richards, *Indigenous Agricultural Revolution: Ecology and Food Production in West Africa* (London: Hutchinson; Boulder, CO: Westview Press, 1985), pp. 55–71; W. E. A. van Beek and P. M. Banga, "The Dogon and their Trees" and J. Pottier and A. Nkundabashaka, "Intolerable Environment: Toward a Cultural Reading of Agrarian Practice and Policy in Rwanda," in E. Croll and D. Parkin (eds.), *Bush Base: Forest Farm: Culture, Environment and Development* (New York: Routledge, 1992), p. 64 and pp. 146–168, respectively; Fairhead and Leach, *Misreading the African Landscape*, pp. 87–88, 90–92, 126–130, 140–142, 194–202; V. Mazzucato and D. Niemeijer, *Rethinking Soil and Water Conservation in a Changing Society: A Case Study in Eastern Burkina Faso* (Wageningen: Wageningen University, 2000), pp. 157–163, 178, table 6.1, and 181–201; and J. Fairhead and M. Leach, *Reframing Deforestation: Global Analysis and Local Realities. Studies in West Africa* (London: Routledge, 1998), p. 149. For Latin America, see E. Graham and D. M. Prendergast, "Maya Urbanism and Ecological Change," in Steen and Tucker, *Changing Tropical Forest*, pp. 102–109; Glaser and Woods, *Amazonian Dark Earths*; Lehmann et al., *Amazonian Dark Earths*.

[59] NAN, NAO 100 Chief Kambonde to NCO, August 8, 1952, and Statement Holongo Amshelelonanda at Ondangwa, July 4, 1952.

bush land and start from scratch.[60] Nigerian farmers also consider soil fertility not as an inherent quality of soils but as a "dynamic characteristic": soils can be improved through the application of organic matter and mounding and ridging.[61]

In 1966, the head of the colonial administration in Ovamboland met with villagers in Ogongo to discuss their removal for the construction of what later became the Ogongo Agricultural College. The farmers insisted that their fields were fertile and well-cultivated and demanded compensation; the colonial government merely had offered them alternative bushland. The farmers argued that it had required great effort on their part to clear the forests, fertilize the soil with manure, and fence their land. They pointed out that they would have to invest heavily in the new land, which they doubted would ever be as productive as the land that they were being asked to abandon. In the face of their persistence, the head of the colonial administration in Ovamboland finally recommended to his superior an exception to the established policy to pay the farmers monetary compensation in addition to giving them an equivalent size plot of land.[62]

The Ogongo farmers successfully had argued that converting a bushland plot into a farm with good soils was a major undertaking. This was especially true where the best sites had been taken up by the first waves of refugees or migrants. Even though the sandy *omufitu* soil that marked the top of the Ovambo floodplain's low ridges was known to be poorly suited for grain cultivation until it had been heavily treated with manure, it became increasingly popular for new farms from the 1940s onward because most of the middle slopes had been occupied.[63] After the field was demarcated and the vegetation was cleared, the soil was literally built up by constructing raised beds or mounds where the millet and other crops would be grown. Manure and other organic matter were carried onto selected fields and fallow cycles also increased soil organic matter.[64]

[60] Wülfhorst, *Moses*, pp. 14–15.

[61] Ishida, Tian, and Wakatsuki, "Indigenous Knowledge and Soil Management," pp. 91–109.

[62] NAN, AHE (BAC) 1/346, Bantu Affairs Commissioner to Chief Bantu Commissioner, Ondangwa, May 16, 1966.

[63] Interviews by the author: Joshua Mutilifa, Omhedi (Namibia), March 8, 1993, and Nahandjo Hailonga, Onamahoka, February 4, 1993.

[64] These technologies are widely used throughout Africa, see, e.g., Richards, *Indigenous Agricultural Revolution*, pp. 55–71; L. C. Gray, "Investing in Soil Quality: Farmer Responses to Land Scarcity in Southwestern Burkina Faso," in Bassett and Crummey, *African Savannas*, pp. 75–76; van Beek and Banga, "The Dogon and Their Trees" and

The raised fields were separated by "paths" that doubled as drainage channels. The mounds varied in size from one to four meters (three to thirteen feet) in length to one to four meters in width, with a height of fifteen to forty centimeters (six to sixteen inches), giving the fields the appearance of a checkerboard or graveyard. An official who had served in Ovamboland for several years in 1935 noted: "in Ovamboland... they use their furrows crisscrossing."[65] Worldwide, mounding was a widely used technique to increase the depth of fertile topsoil by concentrating it on a smaller planting surface, improving root development through better aeration, and protecting crops from water logging and flooding. Significantly, studies on such soil fertility management practices as manuring and mounding stress the heavy labor inputs that are required.[66]

Enhancing soil fertility by applying organic matter required medium- to long-term investment. Depending on their individual circumstances, farmers improved their soils by applying manure, ashes, termite mound earth, and muck and silt that were scooped from water holes and water reservoirs.[67] Cattle manure especially was a critical resource: it was

Pottier and Nkundabashaka, "Intolerable Environment: Toward a Cultural Reading of Agrarian Practice and Policy in Rwanda," p. 64 and pp. 146–168, respectively; Fairhead and Leach, *Misreading the African Landscape*, pp. 87–88, 90–92, 126–130, 140–142, 194–202; Mazzucato and Niemeijer, *Rethinking Soil and Water Conservation*, pp. 181–201.

65 Lea Paulus, interview with the author, Onandjaba (Namibia), June 17, 1993; NAN, NAO 11, Annual Reports Ovamboland FMS 1934 and 1937; A450, 12, SWA Commission: Minutes of Evidence August 12 and 13, 1935, Hahn, p. 651 and vol. 14, Eedes, p. 746; A450 7, Annual Report Ovamboland 1943; NAO 62, Agricultural Report Ovamboland 1953; BAC 133, Agricultural Report Ovamboland 1955/1956. See also Lau, *Carl Hugo Hahn Tagebücher*, July 22–23 and 27, 1857; AVEM, RMG 2599 C/i 19, Beersmann, Omburo, January 6, 1892; NAN, A233, J. Chapman, 1903–1916, 1876[?] [pp. 61–62] and NAO 104 Anderson to Hahn, Jordan diary; Möller, *Journey*, p. 110. The cultivation beds are known as *oimungu* (pl.) and the paths between them as *eendjila* (pl.).

66 See Scoones (ed.), *Dynamics & Diversity*, especially the following contributions: A. Konde et al., "Creating Gardens: The Dynamics of Soil-Fertility Management in Wolayta, Southern Ethiopia," pp. 45–77; I. Dembélé et al., "Seizing New Opportunities: Soil Fertility Management and Diverse Livelihoods in Mali," pp. 78–115; C. Chibudu et al., "Soils, Livelihoods and Agricultural Change: The Management of Soil Fertility in the Communal Lands of Zimbabwe," pp. 116–165.

67 On manure use, see NAN, NAO 62, Agricultural Report Ovamboland, Omafu, November 30, 1953; Kalolina Naholo, interview with the author, Ohamwaala (Namibia), January 26–27, 1993. On ashes, see Kalolina Naholo, interview by the author, Ohamwaala (Namibia), January 26–27, 1993 and NAN, A450, 12, SWA Commission: Minutes of Evidence, Otjikango, August 9, 1935, Gawthorne (CEM), p. 568. See also A450, 7, Annual Report Ovamboland 1937 and NAO 64, Telegram NCO to Commandant Fourie, Ondangwa, September 25, 1953.

essential in creating fields when a village was first settled and it was important for maintaining the soil fertility on fields that were subjected to almost continuous cropping. Colonial agricultural experts emphasized that crop cultivation in Ovamboland was highly dependent on manure. In 1947, the director of the department of agriculture for colonial Namibia wrote:

It must be pointed out that the annual Mohangu [pearl millet] crop is very largely dependent on the supply of manure which by normal standards is very heavily supplied in Ovamboland. As the country is becoming gradually more and more denuded the stock that can be maintained in the cultivated area during the winter will become less and less. Accordingly the supply of manure will also diminish.[68]

A 1958 report estimated that 1 to 1.5 tons of manure was applied per 1.7 ha on a total of 50,000 ha, with a grand total of 100,000 tons per year. In fact, colonial experts at the time considered the amount of available manure to be a major constraint on field size; years later, interviewed farmers similarly underscored the importance of manure to creating and maintaining fields.[69] The 1958 estimate of the amount of manure applied may be on the high side but it does not seem to be totally off the mark. Research in Mali, e.g., suggests an average fecal matter production of three kilograms per day per head of cattle.[70] In 1958, veterinary staff inoculated 333,000 head of cattle in north-central Namibia. Using a conservative figure of 300,000 cattle producing three kilograms of fecal matter per day for 150 days per year in the villages (the cattle were at remote cattle posts the remainder of the year) yields a total fecal matter production of 135,000 ton. Given that cattle were herded in the villages and on the edge of the villages during the day and confined to the on-farm cattle enclosures during the night, half of the fecal matter

[68] NAN, AGR 538, Director of Agriculture Windhoek, October 26, 1949, Agricultural Survey of Ovamboland with Reference to Agricultural and Stock Improvement in that Area.

[69] NAN, NAO 62, Agricultural Report Ovamboland 1953. On manure use AGR 538, Director of Agriculture Windhoek, October 26, 1949, Agricultural Survey of Ovamboland with Reference to Agricultural and Stock Improvement in that Area; BAC 133, Agricultural Report Ovamboland 1956/1957; BAC 133, Quarterly Report Agriculture Ovamboland, June 1958; BOS, Oshikango Stamsecretaris G. Kautwima to Omutonateli Wowilonga, Ohangewena, March 1, 1967. On the restrictions on access to manure by household members, see Joseph Nghudika, interview with the author, Onamahoka, February 3, 1993.

[70] M. S. Dicko-Toure, "Measuring the Secondary Production of Pasture: An Applied Example in the Study of an Extensive Production System in Mali," Le Houérou, *Browse in Africa*, pp. 247–253.

(62,500 ton) was likely to have been deposited on-farm. In addition, because cattle were allowed to graze in the fields for one month after the harvest, the entire production of fecal matter during this period probably was deposited on-farm, which adds another 13,500 tons to the 62,500 tons, for a total of 76,000 tons. The 1958 colonial report presumes that the manure would be evenly distributed across the 50,000 ha of farmland, which would mean an average annual addition of almost 1.2 ton per ha, or roughly halfway between West African estimated applications that range from as low as 0.25 ton per ha to as high as 2.1–3.2 ton per ha. The nutrient contents of manure varies enormously based on livestock species, the season, and the quality of livestock feed, but, in general, the recommended applied quantity of manure to maintain cereal production ranges from 1.2 to 3 ton per ha.[71] Browse high in tannins, e.g., increases nutrient availability because it decreases nitrogen losses in urine and manure. Mopane leaves, which constitute a major source of browse in Ovamboland, are high in tannins.[72]

In terms of cattle manure availability per hectare, Ovamboland farmers were therefore at the lower end of the recommendation if they had distributed the manure equally over the total arable land. Cattle manure, however, was not applied to all the fields on a single farm and it also was not evenly distributed over the fields that did receive manure.[73] Moreover, households would have had varying access to cattle manure because not all households had access to cattle. In addition, cattle manure was just one of several products used to fertilize fields; goat manure, anthill earth, and ashes from the cooking fires were also applied to crop fields. The practice of confining livestock to on-farm kraals during the night facilitated the capture of livestock urine which is high in nitrogen. Urine inclusion in field soils also enhances the ability of crops to absorb other nutrients.[74]

[71] On tanniferous browse and nitrogen retention, see S. A. Tarawali et al., "The Contribution of Livestock to Soil Fertility," in W. A. Dick et al., (eds.), *Sustaining Soil Fertility in West Africa* (Madison, WI: Soil Science Society of America and American Society of Agronomy, 2001), pp. 281–304. On mopane browse in Ovamboland, see Chapter 5.

[72] The original sources provide the cultivated surface in *morgen* with a total surface of 70,000 *morgen* and the application of 1 ton per *morgen*. For the West African estimates, see Tarawali et al., "The Contribution of Livestock to Soil Fertility," pp. 281–304.

[73] Typical figures for Niger show an application of 1.3 tons per ha per year on 30–50% of the fields. See Tarawali et al., "The Contribution of Livestock to Soil Fertility," pp. 281–304.

[74] N, P. and K uptake by crops is enhanced by captured urine, see Tarawali et al., "The Contribution of Livestock to Soil Fertility," pp. 281–304.

Another method that resulted in dark earth involved the practice of the homestead fallow. Ovamboland's farmers moved the family homestead every one to four years and often moved the livestock kraals as well, but they did so *within* the confines of the original clearing and its fence. The new homestead was constructed to the east of the old one, and the old homestead site subsequently became a much-prized field. In the Oukwanyama district, the site that a homestead had been moved from was known by a special name: *elunda*. In 1993, e.g., Julius Abraham lived on the same farm plot that his father had cleared in the 1920s, but his homestead was located across the field from where his father's original homestead had been established. Although the homesteads usually were moved over a short distance, the operation nevertheless was extremely labor intensive. A household might receive assistance from relatives and neighbors for up to two or three days. In the case of royal homesteads in the pre-colonial floodplain, the move could require the labor of up to 100 men. Homestead buildings typically were moved during the end of the dry season, from as early as August to as late as early December. Young men and boys performed most of the hard work, but it was organized and led by older men, who designed the plan for the new homestead and selected the materials.[75] A 1941 case is illustrative. An official found that

[75] Julius Abraham, interview by the author, Olupito, June 16 and 18, 1993; Möller, *Journey*, pp. 112, 126; AVEM 2518, C/h 34, Speiker, Ondjiva, July 19–24, 1906, and RMG 2630 C/k 7, Sckär, "Kurze Geschichte der Ovakuanjama," appendix Sckär to Inspector, Namakunde, October 2, 1912; NAN, A450, vol. 12, SWA Commission: Minutes of Evidence (1935), vol. 9, August 9, 1935, evidence by Rev. Gawthorne (Anglican Mission), p. 569; Anonymous, "Ovambo Customs" (1926), and C. H. L. Hahn, "Preliminary Notes on Certain Customs of the Ovambo," *Journal SWA Scientific Society*, 3 (1927/28), pp. 16–18, and "Property Rights" and "Tribal Laws and the Customs of the Ovambos," and vol. 10, an outline written on blue paper mentioned on p. 2 "[f]ields moved periodically" and "[p]eriodically [the houses are] shifted." See also RCO 4, Ipumbu to NCO, Ukuambi, October 23, 1923; NAO 44, O/C Oshikango to NCO, Oshikango, March 17, 1931 and NCO to Sec. SWA, Ondangwa April 20, 1931; NAO 19, Monthly Reports Ovamboland, June–July 1934; NAO 45, NCO to Sec. SWA, February 17, 1941; NAO 46, Testimony Hamnyela Nashipili, Grootfontein, May 5, 1941, Statements Hamnyela Nashipili, Dirk Jacobus Greyling, Martin Kapenda, Namkoloka Nashipili, Ondangwa, February 12, 1941; NAO 91, Statement Ananias Ruben, Ondangwa, September 4, 1953; NAO 20, Monthly Reports Ovamboland, January–February 1941; NAO 67. The tribal Secretary for the Chief Shetuatha Mbashu (Uukwaluthi) to NCO, Government Camp, Oom Piet, Uukualuuthi, September 22, 1953; BAC 133, Agricultural Officer Ovamboland, Agricultural Report Ovamboland, 1955/1956. In the previously mentioned document "Ovambo Customs," the word for the removal is given as oluthilu. Compare G. W. R. Tobias and B. H. C. Turvey, *English-Kwanyama Dictionary* (Johannesburg: Witwatersrand University Press, 1991 [1954]), p. 118: "*-dulika*: move a kraal to a new site across the garden" and B. H. C. Turvey (comp.) and W. Zimmermann

his investigation of a crime site was hampered because the homestead had just been moved: "[all] that remained of the old kraal... was the outside circle of poles and one grass-roof-shelter... there was a circular space of about 25 to 30 yards diameter, enclosed with poles planted close to one another." In his testimony, the owner, Natanael Tshavura, explained that he had moved the homestead in September 1940 "because it was dilapidated. It is our custom to shift our kraals every 3–4 years to a different part of the field. That evening I had provided beer and food for the people who had been helping me in the morning... about thirty people."[76] The homestead fallow was restricted to the fields of the husband and wife; it did not involve the fields of any junior household members, including junior wives and children.[77] The homestead rotational fallow system was still widely used in the early 1990s: two surveys indicated that 50–80 percent of the respondents practiced it. The frequency of the rotation was highly dependent on the availability of labor.[78]

A substantial percentage of the rich vegetative contents on the former homestead sites that formed the dark earths derived from decaying

and G. B. Taapopi (eds.), *Kwanyama-English Dictionary* (Johannesburg: Witwatersrand University Press, 1977), p. 83: *oludilu*: removal of homestead to a new site. The -th- in the Ondonga orthography is generally a -d- in the Oukwanyama language orthography.

76 NAN, NAO 46, Greyling, Investigation Diary: death of Kasmia Shivuka Hingo, Ondangwa, [1941] and statement Natanael Tshavura, appendices to NCO to Attorney-General, Ondangwa, February 24, 1941. See also interviews by the author: Nahango Hailonga, Onamahoka, February 4, 1993, and cf. Joseph Nghudika, Onamahoka (Namibia), February 3, 1993; NAN, NAO 19, Monthly Reports Ovamboland, June–July 1934; A450, 9, "Tribal laws and customs of the Ovambos" and "Hahn, Rough notes on tribal customs of Ovambo;" NAO 46, statement Kenatuka Keivanga, March 9, 1939, Rex vs. Chiwetha Hiyanga, Chief Warden Windhoek Gaol to NCO, Windhoek, November 23, 1942.

77 Joseph Nghudika, interview with the author, Onamahoka (Namibia), February 3, 1993.

78 NAN, BAC 133, Agricultural Reports Ovamboland 1955–1956 and 1956–1957. Philippus Haidima mentioned the lack of poles as a restraint in moving his homestead more regularly; when he last moved his homestead in 1990, he obtained some replacement poles from just across the border in Angola, interview by the author, Odibo (Namibia), December 9, 1992. The last time Kaulikalelwa Oshitina Muhonghwo moved her homestead was ten years ago, Kaulikalelwa Oshitina Muhonghwo, interview by the author, Ondaanya (Namibia), February 1, 1993. On shifting homesteads and the fertility of the site, see also interviews by the author: Moses Kakoto, Okongo (Namibia), February 17, 1993; Joseph Kambangula Omboloka (Namibia), February 25, 1993; Philippus Haidima, Odibo (Namibia), December 9, 1992; and Joseph Nghudika, Onamahoka (Namibia), February 3, 1993. OMITI A5.2.8-A5.2.11; Erkkilä's survey indicated that half of the respondents praticed homestead rotation in the early 1990s, see Erkkilä, "Living on the Land," pp. 92–95; NAN, BAC 133 f. HN 8/21/4/1, Agricultural Officer to NCO, "Report of travel," Ondangwa, July 4, 1956, and Agricultural Report Ovamboland 1956/1957.

woody materials, including the palisade and hut poles. On the basis of measurements taken on six farms in the early 1990s, the construction of a single north-central Namibian indigenous farm may have required 55 cubic meters of wood, including thirty-five cubic meters for the palisade. The oven-dried weight of 55 cubic meters of wood is approximately 18 tons. If half of the wood had to be replaced every ten years because of rot and termite damage, the process contributed almost one ton of carbon-rich organic matter per year to the formation of on-farm black earths, with at least half of the amount of wood being sourced off-farm.[79] Many farms may have been substantially smaller, however, with a smaller contribution to on-farm dark earth organic matter from decaying wood. The palisades of the sample homesteads used in the above estimates s contained 865–2,720 poles each, but the palisades of forty-two out of forty-nine homesteads in a 1967 sample ranged from approximately 250 to 1,000 poles, suggesting a palisade wood use that is two-thirds lower than the thirty-five cubic meters in the sample from the early 1990s. In addition, many households did not have cattle and thus would have had a more modest livestock kraal for their smallstock. A more representative, smaller homestead therefore may have used twelve cubic meters of wood for the palisade, ten cubic meters for the farm fence, five cubic meters for the livestock kraal, and five cubic meters for the huts. A total of thirty-two cubic meters of wood, with a 50 percent replacement rate per decade, would suggest sixteen cubic meters or five tons of organic matter production per decade, or half a ton per year, with at least a quarter ton of net gain because of off-farm wood harvesting.[80] The quarter ton of net gain from off-farm woody vegetation, of course, translates into a direct environmental loss for the source site. According to the 1991 national census, 90 percent of the 100,000 households counted in north-central Namibia lived in "traditional" homesteads and would have contributed to both on-farm dark earthing and off-farm nutrient drainage.[81] Although alternative farm construction materials were increasingly used, including wire fencing, corrugated iron, and clay and cement bricks, wood remained an

[79] On the replacement rate of construction wood, see Chapter 4. For the quantity of wood used for a farm in the 1990s, see Erkkilä, "Living on the Land," pp. 51, 100, table 12 and Erkkilä and Siiskonen, *Forestry in Namibia*, p. 154.

[80] On the basis of figures derived from NAN, OVA 53, Sec. SWA to Sec. Agriculture Owambo, Windhoek, June 24, 1974, appendices A–C.

[81] Namibia, 1991 Census, Report A, Statistical Tables, vol. v, table H01 and E. Kreike, "*Architects of Nature: Environmental Infrastructure and the Nature–Culture Dichotomy*" (Ph.D. Thesis, Wageningen University, 2006) pp. 97–98.

PHOTO 2. Making dark earth: heaps of manure and ashes spread on the fields. The palisaded homestead is to the left (photo by the author, 1992).

important construction material and the process of on-farm dark earth formation had not ceased by the close of the twentieth century.

The contribution of decaying construction wood was also more modest for households with alternative construction materials. Seven of the forty-nine farms in the 1967 sample had grain stalk palisades from 50 to 150 meters long. In a 1993 sample survey, 4 percent used grain stalks for the palisade or huts. Grain stalks had to be replaced every year and their decay added substantially to black earth formation, but the stalks probably originated largely on-farm, resulting in a much-reduced net on-farm soil nutrient import.[82] Decomposing sun-dried bricks made from termite hill clay also added to the in-field dark earth formation process. Missionaries introduced the use of sun-dried bricks for constructing huts and homes early in the twentieth century. The use spread throughout north-central Namibia. In two sample surveys of more than 200 households in 1967 and 1993, respectively, half and two-thirds of the farms contained at least one clay brick structure.[83]

As mentioned previously, not all fields were fertilized and not all the fields that received organic matter were fertilized equally. Until the 1950s,

[82] NAN, OVA 53, Sec. SWA to Sec. Agriculture Owambo, Windhoek, June 24, 1974, Appendices A-C and OMITI 4.3.11 (1993 sample). See also Kreike, "Architects of Nature," pp. 23–24 and 96.

[83] For the 1967 sample, see NAN, OVA 53, Sec. SWA to Sec. Agriculture Owambo, Windhoek, June 24, 1974, appendices A–C. For 1993, see OMITI 4.3.11.

the arable land of individual farms in the Ondonga and Oukwanyama districts was subdivided. In Oukwanyama in the 1920s and 1930s, after being allocated a farm, the husband and the wife divided the cleared plot in two parts: one for the husband and one for the wife. In polygamous households, each of the wives received a field, but not at the expense of the share that had been allocated to the first or "great wife." The homestead was rotated between the fields of the husband and the first wife; but it was not rotated to the fields of the co-wives.[84] In addition, fields could be allotted by the owner(s) of the farm to individual members of the household, including children, and to recently married men and their spouses who often lived with the husband's parents before the couple received a farm of their own.[85] Junior members of the households who had been allocated fields, however, rarely had access to manure.[86] Social status thus contributed in important ways to an individual's ability to create and maintain a fertile soil-based environmental infrastructure.

In any given year, people cleared more or less of their plot, depending on the rains and the available labor. Uncultivated land remained or reverted to a bush fallow. Some agricultural reports on Ovamboland provide estimates of farm fallow. For example, of the 71,961 ha of fields available in 1957, 2,878 ha were not cultivated because of the late rains, although the fields had been cultivated in the previous year. In the 1978/1979 season, 40,000 ha of a total of 190,000 ha of "plowland" were not cultivated, despite having been cultivated during the previous rainy season.[87]

[84] Joseph Nghudika, interview by the author, Onamahoka (Namibia), February 3, 1993; Kreike, *Recreating Eden*, chapter 6.

[85] Kreike, *Recreating Eden*, chapter 6 and NAN, NAO 91, statement Titus Muatelai Kakonda, Ondangwa, May 10, 1954. Sub-allocating fields to individual household members seems to have been quite common. See, e.g., M. J. Watts, "Idioms of Land and Labor: Producing Politics and Rice in Senegambia," in T. J. Bassett and D. E. Crummey (eds.), *Land in African Agrarian Systems* (Madison: University of Wisconsin Press, 1993), pp. 157–193, especially 176.

[86] NAN, NAO 62, Agricultural Report Ovamboland 1953. On manure use, see AGR 538, Director of Agriculture Windhoek, October 26, 1949, Agricultural Survey of Ovamboland with Reference to Agricultural and Stock Improvement in that Area; BAC 133, Agricultural Report Ovamboland 1956/1957; BAC 133, Quarterly Report Agriculture Ovamboland for the Quarter ending June 30, 1958; BOS, Oshikango Stamsecretaris G. Kautwima to Omutonateli Wowilonga, Ohangewena, March 1, 1967. On the restrictions on access to manure by household members, see Joseph Nghudika, interview with the author, Onamahoka (Namibia), February 3, 1993.

[87] See NAN, NAO 103, Census of Agriculture Ovamboland 1949/1950; BAC 133, Agricultural Report Ovamboland 1956/1957 and Quarterly Report Agriculture Ovamboland June 1958; AHE (BAC) 1/352, Annual Reports Agriculture Ovamboland, 1966 and 1968; OVA 6, Annual Report Agriculture Ovamboland 1978/1980.

The farm bush fallow had multiple uses: it served as reserve browse or cropland as well as a source of wood and livestock fodder. It also served as the household's lavatory and a reservoir of night soil.[88] Having an expansive bush fallow, however, may have been discouraged, given the colonial administration's urgings to make larger fields even as arable land grew scarcer. For example, in the early 1950s, when Nuthano Shanijengana, who was old and frail, left part of her large farm in Omakango in Ondonga district under fallow, the village headmen used it as an excuse to take away half of her land.[89]

Every year, households cleared crop residue, bush, and other vegetation from the area that was intended for cultivation before the onset of the rains, planting the crops after the first good rains had fallen.[90] The removed vegetation, including crop residue and small bushes, was collected in heaps in the field. After the debris had dried, farmers burned it and spread the nutrient-rich ashes over the fields.[91]

Dark Earth and Raised Fields: Drainage and Irrigation

The phenomenon of raised fields is little studied. Such fields are considered to be labor-intensive features of wetter regions, including river floodplains. Because they serve to drain excess water from the crops, raised fields are thought to be rare in semiarid and arid regions.[92] The environmental challenge in semiarid and arid regions, however, is not merely water deficiency. Typically, rainfall occurs in highly concentrated and intense rainfall events, resulting in flash flooding, a danger that looms large in ephemeral river valleys in both arid and semiarid regions across

[88] Nahango Hailonga, interview with the author, Onamahoka (Namibia), February 4, 1993 and NAN, NAO 36, Annual Health Report 1937.

[89] NAN, NAO 100, Chief Kambonde to NCO, Okaloko, January 21, 1954, and NCO to Chief Kambonde, [Ondangwa], January 19, 1954. See also NAO 19, Monthly Report Ovamboland, June 1933 and A450, 7, Annual Report Ovamboland 1937.

[90] For example, in September of 1925, 1926, and 1932, people were reportedly clearing their fields in anticipation of the rains, and in December 1924 and 1927 after the rains had commenced they were cultivating and planting crops, NAN, NAO 18–19, Monthly Reports Ovamboland, December 1924, September 1925 and 1926, October and December 1927, September–October 1932; NAO 89, NCO to Ikasha Kiyala, Ondangwa, October 28, 1948.

[91] Kalolina Naholo, interview by the author, Ohamwaala (Namibia), January 26–27, 1993, and NAN, A450, 12, SWA Commission: Minutes of Evidence, Otjikango, August 9, 1935, Gawthorne (CEM), p. 568. See also A450, 7, Annual Report Ovamboland 1937 and NAO 64, Telegram NCO to Commandant Fourie, Ondangwa, September 25, 1953.

[92] Denevan, *Cultivated Landscapes of Native Amazonia*, pp. 214–290.

the globe. In Ovamboland's semiarid environment, rainfall is associated with convectional storms, resulting in heavy precipitation in a very short time frame. In 1937, Native Commissioner Hahn commented on the occurrence of a very *unusual* drizzle: "[a] notable feature of this season[']s rain is the soft and almost mist-like precipitation lasting for hours at the time. Usually rain in this country is heavy and tropical accompanied by much thunder and lightning."[93]

Local precipitation and run-off from the highlands to the north caused the rapid flooding of the lower areas of the Ovambo floodplain. In 1886, a missionary on the Angolan side of the border saw the water in the flood channels rise to waist deep in just two weeks. In 1925, one of at least three successive floodwater events south of the border increased water levels in the channels by eight inches (20 cm).[94] Flooding is facilitated by the presence of soils of low permeability, especially in the low-lying flood channels of the floodplain, and in the abundant depressions or pans. Even the higher, more permeable sandy ridges (or "dunes") between the flood channels may see flooding during intense rainfall. Their sandy and loamy top soils are often thin and overlay less permeable subsoil layers, including clay and calcrete. In northern floodplain Ombadja in the early 1900s, e.g., the middle slopes of the ridges between the flood channels were not used for crop cultivation because they became easily waterlogged.[95] A visiting agricultural officer in 1947 noted that "[t]he soil in Ovamboland is by no means well adapted for Agriculture as the surface soil from 4″–6″ [10–15 centimeters] is sand without any body and below that it is mostly a stiff inert clay."[96]

The danger of flooding in crop fields was countered by the use of raised fields. Farms stretched more or less perpendicular from the top of the ridges downward toward the flood channels. The top of the ridges consisted of deep sandy soils with tree and bush vegetation. Farmers

93 NAN, NAO 20, Monthly Reports Ovamboland, January–February 1937.

94 Kreike, *Re-Creating Eden*, p. 22 and NAN, NAO 18, Monthly Report Ovamboland, May 1925

95 Lima, *A Campanha*, pp. 14–18, 111–122, 175–183, 200–203; CNDIH, Avulsos, Caixa 3703, Processo Missão de Estudos no Sul de Angola, Relatório do Mez de Outubro [1914] (this report ascribed the shrub size of woody vegetation to aridity) and Caixa 4130, Governo do Distrito 31–20.1, Relatório sobre a Occupação, Lubango May 22, 1909; Petrus Shanika Hipetwa, interview by the author, Oshiteyatemo (Namibia), June 17, 1993.

96 NAN, NAO 101, H. E. Melle [Assistant Director Agriculture] to the Director of Native Agriculture Pretoria, August 6, 1947. See also WAT 144, Dr. H. Martin (Department of Works), "Ovamboland: Geology and Groundwater Conditions," November 5, 1947.

selected the middle and lower slopes of the ridges as fields and intersected the raised fields by "paths" that functioned as drainage channels. Native Commissioner Hahn provided this description in 1924: "[t]he whole country [Ovamboland] really consists of a net work of watercourses.... In the rainy season these become flooded, giving Ovamboland the appearance of a chain of islands. On the edges of these islands Ovambos build their kraals and their gardens."[97] During the dry season and drought years, the fields closer to the flood channel's banks have more humid soils; these same fields, however, are quite vulnerable to flooding during heavy rain. Upslope fields experience less percolation from standing water in the flood channels and are also less subject to flooding. Fields in or near pans are subject to similar dynamics. As the population density increased in the floodplain during the twentieth century, fields adjacent to or even in the flood channels or pans became more common, making for very high-risk farming.[98]

Floods in fields and the resulting crop destruction were a danger in any year of heavy local rainfall, especially when combined with heavy rains upstream on the Angolan side of the floodplain. In March 1925, the rising water in the flood channels inundated fields, ruined the standing crops, and flooded homesteads in the Oukwanyama district. In Uukwambi district to the southwest of Oukwanyama, 75 percent of the fields were flooded and ruined.[99] In Oukwanyama, the homestead as a rule was situated in the center of the farm plot, on the middle slope. That homesteads flooded in March 1925 suggests that the affected households might have built their farms and homesteads further downward on the slopes than was advisable. The influx of a massive refugee exodus from the Portuguese side of the border in the late 1910s and early 1920s forced people to create their farms in less desirable locations.[100] Alternatively, one or more of the March rainfall events may have been so heavy that some of the mid-slope soils became saturated. The Uukwambi and Ondonga districts had an even flatter topography than Oukwanyama, making fields more vulnerable to flooding. A January 1927 report noted the expectation of a "fairly good crop... although a great many fields in the Ondonga

97 NAN, NAO 18, Hahn, Notes on Ovamboland for the Administrator (of SWA), Windhoek, May 15, 1924.

98 See, e.g., Moses Mundjele, interview by the author, Onengali (Namibia), May 17, 1993.

99 NAN, NAO 18, Monthly Reports Ovamboland, March–April 1925. The flood is known as the *efundja*.

100 Kreike, *Re-Creating Eden*, pp. 52–80.

[district] have been ruined by inundation althoguh [*sic*: although] the rainfall has not been excessive."[101] Heavy rains in February 1934 saturated the soil when the grain crops were standing up to six to eight feet (1.8 – 2.4 meters) high in places,

> and there is now so much standing water about that many fields are unfortunately swamped. Natives whose fields are situated on higher land will however ... reap a good harvest. A spell of dry weather set in about 8 days ago and this has been of great benefit to crops which are now beginning to ripen. It is also drying up the soil which in parts is much too soggy to allow the corn to mature. This position exists in all the tribal areas. Melons, beans, calabashes, etc., are plentiful.[102]

The inundations made the use of motorcars impossible and a hapless Native Commissioner Hahn had to rely on an ox wagon to assess the impact of the excessive rains on the crops:

> [c]lose enquiries and observations were made in regard to the natives' food supply.... I find, however, that the hardy Omuhongo [mahango millet] and Kaffir Corn [sorghum] have withstood these unusual conditions [excessive rains and the sodden ground] remarkably well and matured fairly well as soon as dry and sunny weather set in and that the natives will reap a fair harvest. Kaffir Corn is plentiful.[103]

Only after the completion of the 1934 harvest in June–July, however, did the extent of the flood damage become clear:

> [o]wing to the excessive rainfall a lot of corn did not mature but they have nevertheless reaped a fairly good harvest, especially in Ondonga and Ukuanayama. The people who suffered most damage were those whose fields were near to watercourses down which flood water flowed. On the whole, however, food is plentiful this year, especially Kaffir Corn.[104]

In 1937, heavy flood waters from Angola again filled north-central Namibia's flood channels to the brim and saturated the soils. Colonial officials were anxious: "[c]ontinued good rains have fallen.... Native crops are thriving wonderfully well everywhere and *unless the lower lying fields become swamped* the tribes will all reap a good harvest." Fortunately, their worst fears did not materialize.[105] Sorghum, a major but secondary crop, fared much better than the millet staple on waterlogged

[101] NAN, NAO 18, Monthly Report Ovamboland, January 1927.
[102] NAN, NAO 19, Monthly Report Ovamboland, January–February 1934.
[103] NAN, NAO 19, Monthly Reports Ovamboland, April–May 1934.
[104] NAN, NAO 19, Monthly Reports Ovamboland, June–July 1934.
[105] NAN, NAO 20, Monthly Reports Ovamboland, January–February 1937. Italics added by the author.

fields. In 1944 and 1950, flooded fields resulted in poor millet harvests, but sorghum production suffered little.[106]

No major destructive flooding events are reported for nearly a decade after 1934. Heavy rains in 1944 broke the pattern. From 400 to 625 mm (16–25 inches) of rain was measured at various points in the month of January alone, and at one of the rain measuring stations the total rainfall for the year was gauged at 760 mm. Floodwaters from Angola flowed all the way to Etosha Pan in the far south of the Ovambo floodplain. Etosha Pan had not been as full in ten years. Roads and tracks became inundated and communications were cut off: "[i]n certain areas there was so much water that even visits on foot became impossible." Rain and floodwater inundated farms and destroyed standing crops: "[h]undreds of native kraals had to be removed as the ground upon which they stood became submerged. The crops too suffered considerably and the Ovambos will not reap anything like a normal harvest of their staple food, 'Omahango [millet].' Kaffircorn crops [sorghum] are somewhat better." In Ondonga district, flooding badly damaged "many of the cornfields of Ovambo" especially in the southern parts where the rainfall had been heaviest. The Finnish Mission stressed that "[m]any people had little harvest, or nothing at all."[107]

During the early 1950s, heavy rainfall again trumped drought as a threat to crops. Heavy rains fell locally and upstream in the floodplain in Angola,

> with the result that the whole of Ovamboland became inundated in March. The standing crops are in great danger. The following note was received from an Ukuambi Headman: 'Much rain has fallen in Ukuambi country. On the 20th Instant [i.e. March] a large gale destroyed our crops and broke down the trees.' The following is another note received from a Messenger in Ongandjera: 'The corn baskets have fallen down. Corn is rotten and wasted. Fishes and sardines can be seen in the camp and in the corn depot.'[108]

The rains continued until late April, causing further damage to the staple: "[t]he crops of millet are poor throughout Ovamboland, having been considerably damaged by the heavy continuous rains and flood waters from Angola. Crops of kaffircorn [sorghum], melons, beans, etc., are

[106] NAN, NAO 11, Annual Report Finnish Mission Ovamboland 1944 and NAO 21, Quarterly Report Ovamboland, January–March 1944; NAO 60, Quarterly Reports Ovamboland, January–June 1950.

[107] NAN, NAO 21, Quarterly Report Ovamboland, January–March 1944 and NAO 11, Annual Report Finnish Mission Ovamboland 1944.

[108] NAN, NAO 60, Quarterly Reports Ovamboland, January–June 1950.

good and the Natives will have sufficient supplies to carry them through to the next harvest."[109] Heavy rains also caused crop damage during the 1953/1954 rainy season. Initially, it appeared that the impact was limited to the western Ongandjera district; after the harvest, however, it became clear that a combination of water and bird predation had caused considerable losses to the millet crop throughout north-central Namibia. The losses were severe enough that most villagers were forced to pay their annual tax in forms other than the customary payment in grain.[110]

The use of raised fields with narrow drainage channels to divert excess water prevented even worse flooding. In 1935, Harold Eedes, the Native Commissioner for the Okavango who had previously served in Ovamboland, lauded the effective field drainage system in use in north-central Namibia. Asked to compare the lack of drainage in Okavango with the on-field drainage system in north-central Namibia, Eedes explained: "[y]ou have seen in Ovamboland that they use their furrows cris[s]crossing. These Okavangos hack their ground just as it is and a big shower comes along and washes all the so-called irrigation away. The whole trouble is that so far as irrigation is concerned they have no system at all."[111]

In 1947, the Assistant Director of Native Agriculture for South Africa highlighted the importance of raised fields in north-central Namibia for flood prevention:

> [t]hough it may be only at intervals of three to four years that the Oshanas [flood channels] bring down floods sufficient to fill the Etosha Pan, in most years they carry enough water to permit the Ovambos to make use of their broad [raised] beds as the waters recede, or to cultivate the neighbouring plots. Practically all the preparation of the seed beds is done by hoe, owing to the presence of inert soil so close to the surface. The Ovambos too are most painstaking in the cultivation of their crops. In a wet season, like the present one, each plant is mounded up to avoid drowning and to obtain an even stand where plants are crowded individual plants of millet and kaffir corn [sorghum] are dug up and transplanted.[112]

The raised fields were restored each year before the onset of the rainy season, a task that occupied much of the month of September. The September

[109] NAN, NAO 60, Quarterly Reports Ovamboland, January–June 1950.

[110] NAN, NAO 61, Quarterly Reports Ovamboland, April–September 1954.

[111] NAN, A450, 12, SWA Commission: Minutes of Evidence (1935), vol. 14, Grootfontein, August 17, 1935, Evidence H. L. P. Eedes, p. 746.

[112] NAN, NAO 101, H. E. Melle to the Director of Native Agriculture Pretoria, August 6, 1947.

1926 monthly report for Ovamboland stated: "[n]atives are commencing to prepare ground for crops in anticipation of the rains." Once the rainy season began, the prepared raised fields were sown first. As the rainy season progressed, additional fields were raised and subsequently sowed. The September–October 1932 report read: "[n]atives have all cleared and prepared their fields which are now ready for planting when seasonal rains have set in. To date .3 inches rain have been registered." The author of the 1935 November–December report wrote: "natives are busily occupied preparing their fields and cultivating their crops."[113]

The raised fields practice was so effective in the face of flooding that it appears to have been a major factor in delaying the widespread adoption of the animal-drawn plow in Ovamboland until well after World War II. In 1957, the agricultural officer even openly questioned whether plowing was really an improvement on Ovamboland's "traditional" raised field cultivation:

> the raised bed cultivation methods, as they are practiced by the Ovambo, give more satisfactory results than the common plow in the fields. Therefore different tools and cultivation methods should be tested in order to create more favorable growth conditions for the crops.[114]

Indeed, according to the agricultural officer, one reason for the upsurge in the use of plows by the early-1950s was that Ovamboland's farmers had developed a technique for using the plow to make raised fields. He noted that only in some parts of eastern Oukwanyama had plowing resulted in the discontinued use of raised fields.[115] A later agricultural report similarly noted that the plow saved a lot of labor when it was used to make raised fields, although the report also added that the cultivation of raised fields could be replaced by better methods.[116] Throughout the 1950s, however, most farmers continued to rely on the hoe and raised fields, especially because hoes remained an essential tool for weeding.[117] Plowing and abandoning the use of raised beds increased the danger of crop flooding:

[113] NAN, NAO 18–19, Monthly Reports Ovamboland, September 1925, September 1926, September–October 1932, November–December 1935; NAO 60–61, Quarterly Reports Ovamboland April–September 1948, October–December 1953.

[114] NAN, BAC 133, Agricultural Report Ovamboland 1956/1957 and BAC 132, Agricultural Officer Ovamboland to NCO, Ondangwa, March 1, 1957.

[115] NAN, NAO 62, Agricultural Report Ovamboland 1953.

[116] NAN, BAC 133, Agricultural Report Ovamboland 1955/1956.

[117] NAN, NAO 62, Agricultural Report Ovamboland 1953 and BAC 133, Agricultural Report Ovamboland 1956/1957.

> in Ovambo[land] the natives only use the plow on higher lands that do not flood so easily in years with exceptionally heavy rainfall. On such fields, the results to date have been quite satisfactory. It can be expected that plowed fields may do very badly if another very good rainy season is experienced.[118]

Flooded crop fields predated the introduction of the plow, but incidents of flooding may have increased with its widespread adoption in north-central Namibia. One reason was that the use of raised fields became less common in some regions. A more important factor was the scarcity of arable land and the adoption of plow technology after World War II combined to expand crop cultivation into the lower-lying and more clayish soils closer to the seasonal flood channels, seasonal rivers, and shallow seasonal lakes or pans. These low-lying fields on soils of low permeability were much more prone to inundation during the rainy season than the mid-slope fields that were farmers' first choice before World War II. Constructing and maintaining raised fields that offered protection against flooding rendered people's lives and livelihoods more secure.

Dark Earths: Environmental Infrastructure and Soil History

Creating environmental infrastructure required the transfer of organic matter to farm and field sites and major investments in labor and other precious resources, and, in the process, transformed the environment: the 1920s and 1930s floodplain was constituted not of (pristine) Nature but of microenvironments sheltered by huts, palisades, and fences, allowing, e.g., the growth of tomatoes, tobacco, and fruit trees. Naturally poor soils that were subject to flooding were chemically and mechanically re-shaped to facilitate sustained crop cultivation, creating dark soils. But the creation and maintenance of dark earths came at a price: it required high and continuous investments and, even though it made farm sites more fertile, the practice drained nutrients from off-farm locations.

The efforts of north-central Namibia's farmers did not result in an environment where Culture dominated Nature or where Culture and Nature existed in harmony. Despite fencing and raised fields, animal predation and flooding periodically caused severe crop losses. Although environmental infrastructure contributed to food security through producing and maintaining fertile dark earths, it did not confer immunity to the vagrancies of either Nature, the Supernatural, or (Hu)Man in the

118 NAN, BAC 133, Agricultural Report Ovamboland 1956/1957.

form of drought and war. In Zimbabwe, e.g., indigenous farmers argued that soil fertility was not entirely within the realm of human control: hard work and knowledge had to be combined with spiritual benefaction through protective medicines, rain prayers, and rainmaking ceremonies.[119]

The history of dark earth production in Africa invites more research not only because it has been so little studied, but also because it offers the possibility for a new perspective on Africa's environmental history that goes beyond declinist or inclinist generalizations.[120] The environing processes detailed for Ovamboland are likely to have been widespread in Africa; manuring, mounding, and ridging, e.g., were common practices throughout colonial and postcolonial Africa.[121] Africa abounds in dark earths, including the clayish vertisols that constitute a challenge for farmers because they are highly fertile, but very prone to flooding, and they require adequate drainage before they can be used for agriculture. Ancient Ethiopia's capital, Aksum, was surrounded by black vertisol and McCann speculates that a drainage system must have been in place to

[119] See Chibudu et al., "Soils, Livelihoods and Agricultural Change," pp. 116–165 (especially 159).

[120] See Showers, *Imperial Gullies*. Reij, Scoones, and others have also made major contributions to historicizing African soils. See Reij, Scoones, and Toulmin (eds.), *Sustaining the Soil*; I. Scoones et al., *Hazards and Opportunities: Farming Livelihoods and Dryland Africa: Lessons from Zimbabwe* (London: Zed, 1996); and Scoones (ed.), *Dynamics & Diversity*. By the early 1980s only 7.5% of Africa's soils had been surveyed, but, the fact did not make colonial or postcolonial experts temper their crisis narratives. On the lack of soil surveys, see R. Dudal et al., "Rice Soils for Food Production," in Institute of Soil Science, Academica Sinica, *Proceedings of Symposium on Paddy Soil* (Beijing: Science Press, 1981), pp. 31–41 (especially 34). Lindert argues that the conventional wisdom about the extent of global soil erosion is grossly overdrawn. He uses historical data (soil profiles) to demonstrate that for twentieth-century China and Indonesia the claims about massive soil erosion, salinization, and desertification claims are myths; if anything, the quality of agricultural soils in both countries improved since the 1930s–1950s. He identifies Africa as the continent where soil research is of the highest priority. See Lindert, *Shifting Ground*. On soil erosion in Africa, see also J. P. Harroy, *Afrique Terre qui Meurt: La Dégradation des Sols Africains sous l'Influence de la Colonisation* (Brussels: Marcel Hayez, 1949).

[121] On mechanical soil management in Africa, see Reij, Scoones, and Toulmin (eds.), *Sustaining the Soil*; Scoones (ed.), *Dynamics & Diversity*; R. Soper, *The Terrace Builders of Nyanga* (Harare: Weaver Press, 2006); and A. Ayers, "Indigenous Soil and Water Conservation in Djenné, Mali," in D. M. Warren, L. J. Slikkerveer, and D. Brokensha (eds.), *The Cultural Dimension of Development: Indigenous Knowledge Systems* (London: Intermediate Technologies, 1995), pp. 371–384. On the use of these practices more generally, see H. David Thurston and Joanne M. Parker, "Raised Beds and Plant Disease Management," in D. M. Warren, L. J. Slikkerveer, and D. Brokensha (eds.), *The Cultural Dimension of Development*, pp. 140–146.

PHOTO 3. Raised fields: the checkerboard appearance of the old-fashioned raised fields with their crisscrossing drainage channels is clearly visible in the center of the image. Just above it is a recently abandoned homestead site marked by a large tree. The farm fence is visible as a dark line through the picture with fields dotted with large trees within the fence and bush vegetation beyond it (photo by the author, 1991).

unlock the area's rich crop cultivation potential and feed the city's large population.[122] In the context of the raised fields, mounding, and ridging employed by the inhabitants of north-central Namibia to successfully cultivate the water-logged soils of the Ovambo floodplain, and the fact that these methods were ubiquitous across much of Africa, McCann's idea is promising. The history of "dark earthing" in Ovamboland suggests tantalizing paths for exploration: are the fertile vertisols of Aksum, of the Ethiopian highlands, and indeed of Africa in general, wholly or in part the product of a history of *environing*? And to what extent is that true for other dark earths in Africa and elsewhere? The Holocene was a critical period for vertisol creation in India, e.g., coinciding with the rise and expansion of settled agriculture. A mere coincidence, or was there a causal relationship between the rise of settled agriculture

[122] McCann, *Green Land, Brown Land, Black Land*, pp. 43–45.

and the appearance of the vertisols?[123] Moreover, the processes of creating and maintaining dark earths as environmental infrastructure also offers new perspectives on the role of humans and Nature. Top soil formation involves not only "natural" processes (e.g., erosion) and human agency, but also livestock (as the producers of manure), termites, and other microfauna. Termite earth, e.g., is widely used as a fertilizer and termites play an important role in breaking down the wood of the huts, palisades, and fences that contribute to soil organic matter accumulation.[124]

123 On vertisols in India, see D. K. Pal, T. Bhattacharyya, P. Cahndran, and S. K. Ray, "Tectonics-Climate–linked Natural Soil Degradation and Its Impact in Rainfed Agriculture: Indian Experience," in S. P. Wani, J. Rockström and T. Oweis (eds.), *Rainfed Agriculture: Unlocking the Potential* (Wallingford, UK: CAB International, 2009), pp. 54–72. Vertisols are relatively recent formations that are difficult to till but are very responsive to zero-tillage agriculture. With notable exceptions, including the Ethiopian Highlands, African vertisols are thought to have been largely unused for crop cultivation until the large-scale introduction of animal and mechanized plowing in the twentieth century. See N. Ahmad, "Occurrence and Distribution of Vertisols," A. R. Mermut et al., "Pedogenesis," and D. F. Yule and T. J. Willcocks, "Tillage and Cultural Practices," in N. Ahmad and A. Mermut (eds.), *Vertisols and Technologies for Their Management* (Amsterdam: Elsevier, 1996), pp. 1–41, 43–61, and 261–302, respectively. The black color of vertisols is partly because of the presence of organic matter, including charcoal, which is homogenized with the clay. Dating of vertisol organic matter gave soil ages of 1,100 and 4,000 years at depths of 15–30 cm and 75–90 cm, respectively. See C. E. Coulombe et al., "Mineralogy and Chemistry of Vertisols," in N. Ahmad and A. Mermut (eds.), *Vertisols and Technologies for Their Management*, pp. 115–200 (especially 120–121 and 162–163).

124 On the role of termites, see M. Slingerland and M. Masdewel, "Mulching on the Central Plateau of Burkina Faso: Widespread and Well Adapted to Farmers' Means," in C. Reij, I. Scoones, and C. Toulmin (eds.), *Sustaining the Soil*, pp. 85–89, and Scoones et al., *Hazards and Opportunities*, pp. 103, 119–120. Termitaria in Ovamboland are closely associated with farms and fields and termite earth enriched soil is identified in local indigenous soil classification systems, see Verlinden, Seely, and Hillyer, "Settlement, Trees, and Termites in Central North Namibia," pp. 307–335, and A. E. M. Hillyer, J. F. McDonagh, and A. Verlinden, "Land-Use and Legumes in Northern Namibia: The Value of a Local Classification System," *Agriculture, Ecosystems and Environment*, 17 (2006), pp. 251–256.

4

Water and Woodland Harvesting

Village Environmental Infrastructure

Water and woodland use and management practices in twentieth-century north-central Namibia demonstrate how and why water and woodland resources constitute environmental infrastructure rather than "natural" resources. Historically and currently, north-central Namibia sustains half of the country's population, and the region is densely settled. North-central Namibia, however, is devoid of any natural dry-season surface sources of water. Beyond the Kunene River to the far northwestern margin, the region has no permanent rivers or lakes. All the water courses that run through it, in the floodplain and beyond, are ephemeral and contain only running water for at best a couple of months each year. Moreover, the dense population's demand for wood for construction and fuel is at levels that are considered by most to be far beyond the natural regenerative powers of the local vegetation and could result in deforestation and desertification. The use of alternative cooking fuels (beyond fuel wood) had hardly made any inroads by the close of the twentieth century. More telling, however, is that an average rural household required an estimated fifty to seventy cubic meters of wood for constructing the elaborate homesteads that characterize north-central Namibia; for comparison, a three-bedroom home in Usambara in northern Tanzania uses only 2.4 cubic meters of wood.[1]

Although most dry season water sources were located on-farm, the use and management of dry season stores of water centered on groups

[1] Erkkilä, "Living on the Land," pp. 51, 100; Erkkilä, *Forestry in Namibia*, p. 154; Kessy, *Conservation*, p. 109.

of neighboring households and entire villages. The same was true for the management of the woodlands that serve as a major source of construction and fuel wood.

Again, the creation of village landscapes was not a one-time, permanent undertaking. Refugees from the Portuguese-controlled Ombadja districts of Angola, who resettled south of the new colonial border in the Okalongo area of Namibia during the 1920s and 1930s, encountered overgrown farms and fields, marked by palm and other fruit trees, water holes, and water reservoirs, the ruins of the kingdom of Haudanu that had been abandoned by its inhabitants after a fierce war in the nineteenth century. There were similar signs of ruined settlements elsewhere in the Ovambo floodplain. Abandoned villages were a common occurrence throughout Africa in the nineteenth and early twentieth centuries because of the prevalence of violence that caused populations to flee their villages, resulting in the collapse of the local environmental infrastructure. For example, after the violent suppression of the Maji Maji rebellion in the early 1900s and the subsequent massive depopulation, wide swaths of southern Tanzania turned into tse tse fly-infested bushlands.[2] Outmigration caused the collapse of western Kenya's Il Chamus' dams and irrigation ditches in the late nineteenth and early twentieth centuries. Population decline resulted in the rapid deterioration of terraces, fields, homesteads, and fences in the northern Cameroon highlands in the

[2] On King Haudanu's kingdom, see Kreike, *Re-Creating Eden*, pp. 15–25. On the Maji Maji revolt's environmental impact, see J. Iliffe, *A Modern History of Tanganyika* (Cambridge: Cambridge University Press, 1984 [1979]), pp. 163–202, 270–271; J. L. Giblin, "Trypanosomiasis Control in African History: An Evaded History," *Journal of African History*, 31 (1990), p. 76. Scorched earth warfare in the late-nineteenth-century Sahel zone of West Africa turned thriving savannas into deserts; see Yves Person, *Samori: Une Révolution Dyula* (Dakar: IFAN, 1975), vol. 3, pp. 1738–1739. For other examples, see H. Weiss, *Babban Yunwa: Hunger und Gesellschaft in Nord-Nigeria und der Nachtbarregion in der frühen Kolonialzeit* (Helsinki: Suomen Historiallinen Seura, 1997), pp. 172, 209–212, 233–236, 325; R. Roberts, *Warriors, Merchants, and Slaves: The State and the Economy in the Middle Niger Valley, 1700–1914* (Stanford, CA: Stanford University Press, 1987), pp. 17–20, 76–172; M. Saul and P. Royer, *West African Challenge to Empire: Culture and History in the Volta-Bani Anticolonial War* (Athens: Ohio University Press, 2001), pp. 173–270; S. Ellis, *The Rising of the Red Shawls: A Revolt in Madagascar*, 1895–1899 (Cambridge: Cambridge University Press, 1985), pp. 71, 97, 125, 129; T. C. Weiskel, *French Colonial Rule and the Baule Peoples: Resistance and Collaboration, 1889–1911* (Oxford: Clarendon Press, 1980), pp. 33–69, 124–127, 190–209; J. Lamphear, *The Scattering Time: Turkana Responses to Colonial Rule* (Oxford: Clarendon Press, 1992), pp. 95–98, 138, 165–169, 176–177, 194–198, 276–277; and J. A. de Moor and H. L. Wesseling (eds.), *Imperialism and War: Essays on Colonial Wars in Asia and Africa* (Leiden, the Netherlands: Brill, 1989).

1960s-1980s as too few hands were left to repair and maintain the elaborate environmental infrastructure.[3]

Water Environmental Infrastructure

In north-central Namibia, as elsewhere, water conditions shaped population density and where humans could settle. People established homesteads and fields on the area's sandy ridges because of the frequent flooding of the lower-lying land between the ridges. Human settlement in turn affected the environment because settlers cleared land for farms, cultivated the soil, and introduced livestock to the area. Because water was scarce in the new villages during the dry season, further growth of the village populations could only be sustained by herding the cattle to distant cattleposts for the duration of the dry season to ration the villages' water resources. Environmental conditions, however, not only offered challenges; they also constituted opportunities: the extreme semiaridness of the region made it tse tse fly-free and thus an important cattle-breeding area. Ovamboland's environment therefore was neither hostile nor hospitable by Nature. Moreover, "natural" water conditions did not dictate human settlement and land use. Much like other societies in water-deficient environments in Africa and across the globe, Ovamboland's inhabitants engaged in water harvesting, constructing dams to catch floodwater and runoff and digging wells, water holes, and reservoirs. With few exceptions, however, water harvesting in Africa has been little studied.[4]

[3] For Kenya, see D. Anderson, *Eroding the Commons: The Politics of Ecology in Baringo, Kenya 1890–1963* (Oxford: James Currey, 2002), pp. 31–47. For Cameroon, see Boutrais, *Mbozo-Wazan*, pp. 42–128. Seth emphasizes that labor availability is a critical factor in the maintenance of indigenous water harvesting systems: a labor exodus may lead to their decline, see S. M. Seth, "Human impacts and management issues in arid and semi-arid regions," in I. Simmers (ed.), *Understanding Water in a Dry Environment: Hydrological Processes in Arid and Semi-arid Zones* (Lisse, the Netherlands: Balkema, 2003), pp. 289–341 (especially 311).

[4] The major exception, of course, is the Nile Valley. For sub-Saharan Africa, however, water harvesting has been little studied, perhaps with the exception of some more or less iconic examples of small-scale irrigation societies, including the Shambaa of Tanzania and the Dogon of Mali. Little or no details are known about water harvesting through wells, water holes, and reservoirs. Many contributions in Reij et al. emphasize the importance of water harvesting in different parts of Africa. See, e.g., M. O. El Sammani and S. M. A. Dabloub, "Making the Most of Local Knowledge: Water Harvesting in the Red Sea Hills of Northern Sudan," Y. A. Mohamed, "Drought and the Need to Change: The Expansion of Water Harvesting in Central Dafur, Sudan," A. Kassogué, M. Komota, J. Sagara, and F. Schutgens, "A Measure for Every Site: Traditional SWC

The Ovambo floodplain dominates the western half of the Kuvelai-Oshimolo watershed, which is delineated by the Kunene River in the west, the Kavango (Cubango) River in the northeast and east, and the Etosha Pan in the south. Although the Kuvelai-Oshimolo watershed lacks any permanent rivers, the Ovambo floodplain consists of an inland delta of rain-fed seasonal watercourses that drains into Etosha Lake. The climate is semiarid: on the basis of statistics from the 1940s to 1990s, average annual rainfall in the southern floodplain was between 400 and 600 mm

Technigues on the Dogon Plateau, Mali," A. Kolawole, J. K. Adewumi and P. E. Odo, "*Firki-Masakwa* Cultivation on Borno, North-East Nigeria," D. Millar with R. Ayariga and B. Anamoh, "'Grandfather's Way of Doing' Gender Relations and the Yaba-Itgo System in Upper East Region, Ghana," J. M. Shaka, J. A. Ngailo, and J. M. Wickama, "How Rice Cultivation Became an 'Indigenous' Farming Practise in Maswa District, Tanzania," F. Hiol Hiol et al., "Traditional Soil and Water Conservation Techniques in the Mandara Mountains, Northern Cameroon," in C. Reij, I. Scoones, and C. A. Toulmin (eds.), *Sustaining the Soil*, pp. 28–34, 35–43, 69–79, 90–96, 117–125, 126–133, 191–201, respectively. Contributions in the monumental multivolume *History of Water* are also scarce in terms of offering details of the history of water harvesting in Africa; most chapters about Africa do not offer much detail on actual practices but highlight beliefs about water. See M. Tagseth, "The *Mfongo* Irrigation Systems on the Slopes of Mt. Kilimanjaro, Tanzania," in T. Tvedt and E. Jakobsson (eds.), *A History of Water Volume 1*, pp. 488–506; J. B. Forrest, "Water Resource Administration and Racial Inequality in South West Africa," in R. Coopey and T. Tvedt (eds.), *A History of Water, Volume 2*, pp. 151–171; D. Gordon, "From Sacred Ownership to Colonial Commons: Water Tenure Systems in Central Africa," F. A. Akiwumi, "'Indigenous People Participation': Conflict in Water Use in an African Mining Economy," E. M. Rinne, "'Seeing is Believing': Perceptions of Safe Water in Rural Yoruba," M. McKittrick, "'The Wealth of These Nations': Rain, Rulers and Religion on the Cuvelai Floodplain," C. M. Namafe, "The Lozi Flood Tradition," in T. Tvedt and T. Oestigaard (eds.), *A History of Water Volume 3*, pp.18–37, 49–77, 269–286, 449–469, and 470–487, respectively; T. Saetersdal, "Rain, Snakes and Sex – Making Rain: Rock Art and Rain-Making in Africa and America," in T. Tvedt and T. Oestigaard (eds.), *A History of Water, Series II, Vol. 1*, pp. 378–404. On water harvesting/irrigation in the Kuruman region of South Africa, see Jacobs, *Environment, Power, and Injustice*, pp. 57–69, 131–132, 151–164. On historical water harvesting in Zimbabwe, see Soper, *The Terrace Builders of Nyanga*, especially pp. 53–60. M. Shahin, *Hydrology and Water Resources of Africa* (Dordrecht: Kluwer, 2002) and J. W. N. Tempelhoff (ed.), *African Water Histories: Transdisciplinary Discourses* (Vanderbijlpark, South Africa: Vaal Triangle Faculty, North-West University, 2005) do not include any discussions about "indigenous" water harvesting. The former focuses on hydrology and the latter on the history of colonial and modern water infrastructure, including dams and urban water facilities. On the importance of a better understanding of indigenous water harvesting practices in dry environments, see Seth, "Human impacts and management issues in arid and semi-arid regions," pp. 289–341. See also T. Oweis and A. Hachum, "Water Harvesting for Improved Rainfed Agriculture in the Dry Environments" and P. Pathak et al., "Opportunities for Water Harvesting and Supplemental Irrigation for Improving Rainfed Agriculture in Semi-Arid Areas," in S. P. Wani, J. Rockström, and T. Oweis (eds.), *Rainfed Agriculture*, pp. 164–181 and 197–221, respectively.

(16–24 inches). Ninety-six percent of the rain fell during the months of November to April, but precipitation was heavily concentrated in the three months of January, February, and March.[5] Rainfall from the far north of the watershed, which reaches the delta in the form of a seasonal flood (*efundja*), and local precipitation collect in a maze of watercourses which intersect the floodplain. In years of heavy rains, the flood feeds the Etosha Pan. The eastern half of the watershed contains only a handful of seasonal watercourses. Early in the long dry season, the surface water in the shallow watercourses and pans is gone: consumed by humans, animals, and plants, and evaporated or percolated into the soil.

The missionary Carl Hugo Hahn reached Ondonga on the southeastern edge of the floodplain early in the dry season of 1857. Even before he entered the actual floodplain, he noted an increasing number of water sources. He described dry watercourses marked by regular flooding, water holes, wells, reservoirs, and a large number of small "pans" (shallow seasonal lakes).[6] A Portuguese soldier who participated in the invasion of the Ombadja kingdoms in the far northwest of the floodplain in the early 1900s described the water holes that were the main dry season sources of water in the area:

> When the rain water supplies have dried up they open up some pools or *cacimbas* in locations where there are small lenses of underground water that have been formed through infiltration and where a clay layer prevents the rainwater from percolating to a great depth. These deposits [of water] are fairly frequent, but give little water; just enough for the support of a few families when the consumption is limited to use for food and drink... sometimes these deposits are reached at a depth of 3–4 meters [10–12 feet], producing 30–40 liters [7–10 gallons] in 24 hours.[7]

The explorer Möller described a water hole at Onoholongo in the southern Ovambo floodplain during his 1895–1896 travels:

> [it] is a funnel-shaped depression about 50 meters [150 feet] in diameter, in the middle of which there are two holes about one meter in diameter that human hands have deepened to wells. At the bottom of these the water of the lower layers of the surrounding sand-veld, which sinks down and is stored after each

[5] J. Mendelsohn, S. el Obeid, and C. Roberts, *A Profile of North-Central Namibia* (Windhoek: Gamsberg Macmillan, n.d.), pp. 9–11.

[6] Lau, *Carl Hugo Hahn Diaries*, Part IV 1856–1860, No. 23 Reise zu den Ovandonga, pp. 975–1081. See also AVEM, RMG 2599 C/i 19, Bernsmann, Ondjiva, December 1891 and Omburo, January 6, 1892; Wülfhorst, *Aus den Anfangstagen*, pp. 12–16; and H. Marquardsen, *Angola* (Berlin: Dietrich Reimer, 1920), p. 43.

[7] "Campanha do Cuamato," *Portugal em Africa*, 14 (1907), pp. 443–448.

rain, gathers. In order to water all the oxen six of my men had to stand the one above the other on steps in the well and haul up the water in buckets... before they all had had enough to drink the well was empty and we had to let the water run in during the night.[8]

Another important improved source of water consisted of dams or ponds. Groups of households organized villagers and kings mobilized subjects to deepen natural pans (small depressions where water collected during the rains) and used the excavated earth to build up the banks. Alternatively, people constructed a small dike (or dam) on the edge of a seasonal watercourse. In the latter case, the dam not only gathered water from local rains but also harvested floodwater. Deepened and improved pans were known as *oitenemba* and they were clearly distinguished from natural pans. Some, if not all, of the "pans" that the missionary Hahn described in 1856 may actually have been *oitenemba* or water reservoirs. This human-made rural water infrastructure of water holes (and some wells) and reservoirs supplied the bulk of human and animal needs during the long dry season in the Ovambo floodplain and facilitated the high population densities.[9]

Refugees and migrants placed increased pressure on dry season water resources in north-central Namibia's villages early in the twentieth century. Before colonial occupation in 1915, much of the middle floodplain along the Angolan-Namibian border had been devoid of any substantial human settlement and lacked a water-harvesting infrastructure. The influx of people multiplied the effects of both seasonal and annual drought conditions. The lack of a dry season water infrastructure became abundantly clear during the 1929–1931 Famine of the Dams, which affected all of north-central Namibia. The drought was marked by severe water scarcity, crop failures, and high livestock mortality, and government food aid was necessary to prevent human starvation. A large share of the food aid was distributed through food-for-work projects that expanded the water infrastructure of wells and reservoirs, or "dams" (singular *etale*, plural *omatale*), thus giving rise to the local name for the event: *Ondjala yomatale* "the Famine of the Dams."[10]

[8] Möller, *Journey in Africa*, pp. 142–143.

[9] Marquardsen, *Angola*, p. 97; "Campanha do Cuamato," *Portugal em Africa*, 14 (1907), pp. 443–448; and de Lima, *A Campanha dos Cuamatos*, pp. 136–140.

[10] NAN, NAO 105, Diaries Native Commissioner, Diary 1929, entries January 15 and February 22, 1929; A450, 7, Annual Report Ovamboland 1940. See also Kreike, "Recreating Eden," chapter 5.

The water holes drew on shallow, highly localized aquifers, each of which could support only a small number of people and animals during the dry season. A 1966 report, e.g., noted that a water hole could sustain at most a few households, echoing the limited yield of water holes highlighted in earlier sources. Usually, water holes could be relied on until the beginning of the next rainy season; in drought years, however, they tended to run dry sooner. In addition, elephants at times destroyed water holes, leaving households and entire villages destitute as occurred in Uukwaluthi district in 1953 when a herd of the animals "filled up the water holes with earth so that the poor people have difficulty always in obtaining water [*sic*]."[11]

Data from the 1970s illustrates the use of water holes: of the thirty-three homesteads on lands that were confiscated by the colonial authorities in Ongwediva in 1976–1978, eight had wells or water holes, including two homesteads that each had two sources of water and one that had three sources of water.[12] The 1991 census found that almost 60 percent of north-central Namibia's nearly 100,000 households relied on "wells" (referring to water holes and wells combined) as a source of water.[13] Water holes and local water reservoirs remained in use and even made a virtual comeback during the early 1990s, when, because of severe droughts, villagers repaired old indigenous reservoirs, water holes, and even entire water hole systems or well fields (singular *etambi*) that had fallen into disuse.[14]

In 1993, half of the respondents to the OMITI survey relied exclusively or additionally on water holes. The water holes were high maintenance: the large majority of the households that used water holes emphasized that they periodically needed to be repaired and cleaned. Rarely was the

[11] See NAN NAO 18, Monthly Report Ovamboland, September 1926; A450, 7, Annual Report 1938; WAT 66, Sec. SWA to Sec. Bantu Administration Pretoria, Windhoek, January 21, 1966; NAO 46, Investigation Diary: Murder of young Ovambo Female, Nimungu Iyambo on or about July 24, 194?; NAO 67, Tribal Secretary for Chief Shetuatha Mbashu (Ukwaluthi) to NCO, Government Camp Oom Piet, Uukwaluuthi, September 22, 1953; and interviews by the author: Kaulikalwa Muhonghwo, Ondaanya (Namibia), January 29, 1993 and Joseph Nghudika, Onamahoka (Namibia), February 3, 1993.

[12] NAN, OVA 55, J. P. Mulligan, Department of Agriculture, Payment of Ovambo houses Ongwediva, July 25, 1978, and Sec. Agriculture to Sec. Chief Minister, November 29, 1976.

[13] Namibia, 1991 Census, Report A, Statistical Tables, vol. v, table H06.

[14] Kulaumoni Haifeke, interview by the author, Oshomukwiyu (Namibia), May 11, 1993, and OMITI 3.1.0 and 3.1.1.

water hole considered "communal" property; instead, as a rule, the water holes were owned by individuals.[15]

Wells and Water Holes

In 1993, 40 percent of households sampled in the OMITI survey stated that wells were an important source of water during the dry season. Two-thirds of the respondents who used wells drew their water from private wells; the remaining third used communal wells.[16] Wells (as distinct from water holes) were a relative rarity in Ovamboland until after the 1940s. Asked in 1993 when the well they relied on for their drinking and cooking water had been dug, only 5 percent of those surveyed specified that it predated the respondent's birth. Twenty-five percent of the respondents reported that their well had been dug during their lifetime and more than 55 percent of the respondents claimed that the well concerned had been dug during the past ten years.[17]

Wells tapped deeper aquifers, penetrating the clay and rock strata directly below the generally thin sandy top layers. Piercing the rocky strata required specialized tools (e.g., pick-axes and shovels) and the work was labor intensive, difficult, and hazardous. Constructing a well was a risky investment because of the presence of saline aquifers and the threat of collapse.[18] As a result, during the 1940s through the 1960s, well construction was pioneered by members of the local elite, the missions, or the colonial authorities, who consequently derived management control over these resources, supervising their use and organizing their maintenance and repair.[19] In 1936, the Ovamboland administration claimed that hundreds of wells had been dug under the supervision of its staff.[20] If headmen took the initiative to construct a source of water, it became the property of the community involved. Well technology – next

15 OMITI 3.15, 3.16, 3.17, 3.18, and 3.19. For the private ownership of water holes, see also, e.g., NAN, WWA [WAT?] 637 f. 31/3/1(ii), Department Water Affairs, Report Re. Watersupply Ovamboland Schools, Appendix to Acting Chief Inspector to Director Water Affairs Windhoek, Otjiwarongo, May 13, 1970.

16 OMITI 3.1.1. On well technology, see Kreike, *Re-creating Eden*, chapter 8. Well technology was rapidly appropriated by Ovambo innovators. The senior headman of eastern Oukwanyama, Elia Weyulu, was considered a well pioneer in eastern Ovamboland. See, NAN, AHE/BAC 1/2, Questionnaire on [Headman] Elia Weyulu 11/3/1968.

17 Kreike, *Re-creating Eden*, chapter 9; NAN, RCO 8, RCO to Director Works, Ondangwa, March 29, 1918; and OMITI 3.11 (*N*=170).

18 Kreike, *Re-creating Eden*, chapter 9; NAN A450, 7, Annual Report Ovamboland 1940.

19 NAN, A450, 9, "Property Rights."

20 Paulus Nandenga, interview by the author, Oshomokwiyu (Namibia), March 28, 1993. See also NAN, A450, 13, SWA Commission, "Memo of Reg[ard] Findings of the

to dams and water holes – was critical to colonizing the eastern part of Ovamboland outside of the floodplain environment, especially from the 1940s onward.[21]

Well technology disseminated rapidly. Unlike the colonial dams and boreholes, wells, similar to water holes, principally became an individually controlled resource.[22] For example, during the dry season of 1969/1970, a pastor at Omunholo in Eastern Ovamboland denied teachers and students access to water from his well, perhaps in part to protect his claims to the water source by resisting its designation as a "school well" and thus public or communal property. The pastor had probably seen the writing on the wall. The Department of Water Affairs inspected and increasingly effectively controlled the school wells and water holes. In 1973, the responsibility for school wells equipped with a hand pump was delegated to the individual schools.[23]

In 1993, more than half of households sampled in the OMITI survey relied wholly or partially on water from a well.[24] Slightly more than half of the wells they used had been dug by neighboring households. One-fifth of households dependent on a well had dug their own; in rare cases, the (village) headman or the "government" had dug the well.[25] Two-thirds of the wells used in 1993 contained shafts lined with wood. A small number of wells had a metal, brick, or concrete lining. In addition, a plate sealed off the top of fully one-third of the wells, but only one-eighth of wells were fenced. One-third of users hauled water up with a winch and bucket; hand pumps were rare.[26]

Half of the OMITI respondents emphasized that everyone who used the well contributed to its maintenance and repair, with one out of each seven respondents mentioning that the heads of neighboring households contributed to well maintenance (these households probably also used

Constitutional Commissions in so far as they affect the Administration of Native Affairs 1936."

21 Kreike, *Re-creating Eden*, ch. 9.

22 For government owned wells, see NAN, OVA 33, Senior Technician Republic South Africa to Director Works, Bantu Affairs Ondangwa, Oshakati, September 15, 1970.

23 NAN, WWA [WAT?] 637, Department Water Affairs, Report Watersupply Ovamboland Schools, Appendix to Acting Chief Inspector to Director Water Affairs Windhoek, Otjiwarongo, May 13, 1970. On school wells, see WWA [WAT?] 636, Owambo: Report Wells at Schools and Clinics, Inspection May 22–June 1, 1972, and WWA [WAT?] 637, Director Water Affairs to Chief Director Ovambo Government, n.p., March 16, 1973 and A. H. I. Bon to Pieterse, n.p., October 11, 1972 Contract D. Homateni, Headmaster, Pymaknu, Ongandjera, November 21, 1972.

24 OMITI 3.5 (*N*=324).

25 OMITI 3.10 (*N*=176).

26 OMITI 3.13 (*N*=159).

the well). In addition, almost half of the respondents stated that a well's owner was responsible for maintaining and repairing the well.[27] Of the households that used a well, the overwhelming majority contributed to the well's maintenance, principally in the form of labor, but sometimes in the form of money or food donations.[28]

Preserving and Rationing Water

The person who organized digging a water hole was considered to be the owner of the resource. Often, the owners were individuals and their water holes would be inherited by their heirs.[29] The owner managed the water hole and could restrict its use in times of scarcity. Regular users were expected to participate in maintenance and repairs.[30] If one water hole ran dry after heavy use, people were allowed to draw water from another water hole.[31] Although the ownership of water holes that were located within the fenced area of a farm was often transferred along with the farm, it seems that it could also be transferred separately from the

[27] OMITI 3.6 (*N*=169). The categories "all the people who use the well" and the "heads of neighboring households" probably overlap.

[28] OMITI 3.7 (*N*=182). Of those who specified what they had contributed, 79% mentioned labor, 5% mentioned money, and 5% mentioned food, OMITI 3.8 (*N*=38).

[29] Interviews by the author: Kulaumoni Haifeke, Oshomukwiyu (Namibia), May 11 1993, and Paulus Nandenga, Oshomokwiyu (Namibia), March 28, 1993. On ownership, see also NAN, RCO 10, Manning and Hahn, Preliminary Memo for Ex[pedition] Force [1915 or 1916] and RCO to Staff Officer Union Forces in South West Africa, December 3, 1916; NAO 13, NCO to Sec. SWA, Ondangwa, August 3, 1934; NAO 99, Ndjuluua, Ondangwa, December 12, 1950, Appendix to Chief Kambonde to NCO, Okarolo, December 15, 1950, and Pingan ya Paulus to Master Nakale, November 6, 1950; WAT 66, Sec. SWA to Sec. Bantu Administration Pretoria, Windhoek, January 21, 1966.

[30] See interviews, Kulaumoni Haifeke, Oshomukwiyu (Namibia), May 1, 1993, and Kaulikalwa Muhonghwo, Ondaanya (Namibia), January 29, 1993, and NAN, A450, 9, "Property Rights."

[31] Interviews by the author: Kulaumoni Haifeke, Oshomukwiyu (Namibia), May 11, 1993, and Paulus Nandenga, Oshomokwiyu (Namibia), March 28, 1993. On ownership, see also NAN, RCO 10, Manning and Hahn, Preliminary Memo for Ex[pedition] Force [1915 or 1916] and RCO to Staff Officer Union Forces in South West Africa, December 3, 1916; NAO 13, NCO to Sec. SWA, Ondangwa, August 3, 1934; NAO 99, Ndjuluua, Ondangwa, December 12, 1950, Appendix to Chief Kambonde to NCO, Okarolo, December 15, 1950, and Pingan ya Paulus to Master Nakale, November 6, 1950; WAT 66,Sec. SWA to Sec. Bantu Administration Pretoria, Windhoek, January 21, 1966. Among the Borana, access to wells can depend not only on clan affiliation but can also be based on who contributes to its maintenance and repair. See M. Niamir, "Indigenous Systems of Natural Resource Management among Pastoralists of Arid and Semi-Arid Africa," in D. M. Warren, L. J. Slikkerveer, and D. Brokensha (eds.), *The Cultural Dimension of Development*, pp. 245–257.

land itself. Native Commissioner Hahn concluded that "[t]he water itself is incapable of ownership. The occupier of a waterhole has the right to draw his own requirements before anyone else. Thereafter the neighbors are entitled to draw [water] provided they have assisted in the annual opening up and cleaning of the waterhole after the rains."[32]

When Kalikalelwa Oshitina Muhonghwo was a girl in 1920s Oukwanyama, her stepfather Kalolo Hambwidi owned a water hole in the sandy soil on the edge of their farm. They re-dug and cleaned the water hole every so often, especially in drought years. After carefully clearing off leaves and sticks that could pollute the water, they removed the sand until they reached the water table. Her stepfather organized the re-digging and cleaning of the water hole, inviting the neighbors – all users of the water hole – to join in what started as a work party and ended as a beer party.[33]

Precious water was not only harvested and stored but also carefully preserved and rationed. To preserve village water stores during the dry season, the reservoirs, wells, and water holes were reserved for human and small stock use only. When sources of standing "natural" surface rain and flood water ran out, herdsmen drove the cattle to dry-season forages far beyond the villages, removing a major consumer. The main reason for the cattle transhumance was, therefore, not merely any shortages of grazing and browsing per se, but, instead the need to preserve water resources. The transhumance calendar was primarily shaped by the availability of water supplies at the villages and at the remote cattleposts. Preserving the water stored at the villages, however, came at a heavy price: cattle, one of the villagers' most precious sources of wealth and valuable products (meat, dairy, leather, manure) was removed from the village for the duration of the dry season, precisely at the time that diets were under great pressure and the threat of hunger and disease loomed.

The timing for when the cattle were driven from the villages depended on the rains and how fast surface water ran out as a result of consumption, drainage, and evaporation. In May 1926, e.g., open "natural" sources of water (the flood channels and pans) were drying

[32] NAN, A450, 9, "Property Rights." See also E. Kreike, "Historical Dynamics of Land Tenure in Ovamboland" NEPRU Working Paper No. 2 for the National Land Reform Conference, Windhoek, Namibia, June 1991.

[33] Kalikalelwa Oshitina Muhonghwo, interview by the author, Ondaanya (Namibia), January 29, 1993. In the Kerawan region of The Gambia, maintenance of hand-dug wells was both continuous and expensive; see R. A. Schroeder, *Shady Practices: Agroforestry and Gender Politics in The Gambia* (Berkeley: University of California Press, 1999), p. 69.

up but were sufficient to meet people's and animals' needs because there had been good rains in April. In June, livestock was still in good shape, although "grazing is already becoming poor – [and] all open water and vleis [pans] are now drying up." By September–November, livestock was in poor shape because of a shortage of "grazing" and water, but November rains brought relief: "[s]tock, mostly at outposts, is in poor condition owing to the severe past months. Now that there is plenty of open water in pans... and young grass coming on conditions generally will soon improve." By December, open surface water and grazing in the villages sufficed to allow the return of the cattle herds: "[cattle] have been brought in from outposts [cattle posts] for milking purposes. Stock generally is improving in condition owing to improved grazing... excellent rains having fallen throughout the country."[34]

In August–September 1927, water and grazing in the villages were so scarce that not only cattle, but also donkeys and even smallstock were herded to the remote cattleposts. The subsequent rainy season was poor. Almost as soon as it ended in April 1928, most open surface water had disappeared. By the end of May, only the larger pans contained any water. By August, water and grazing for livestock were in very short supply across Ovamboland, resulting in tensions over access to water. Light September rains had little impact: "stock is in bad condition. Several small showers have... fallen but insufficient [to] leave open water or make any appreciable difference to the grazing." The standing water that remained in the pans after additional rains in November disappeared rapidly and December brought no relief. Not only did scarce water reserves in the villages preclude the return of the cattle, but even water reserves at the cattleposts ran out. In despair, cattle owners and herdsmen drove their animals from the eastern Ovamboland cattleposts to the old cattleposts in Oshimolo, across the border in Angola. Conflicts about water stores erupted at the cattleposts and in the villages because the water stored in water holes and reservoirs no longer sufficed for even the smallstock. Livestock mortality rose sharply: "[h]undreds of cattle have died all over the country from thirst and starvation, and unless rain falls soon most of the [small]stock in the inhabited areas [the villages] will perish."[35]

Only February 1929 rains refilled the pans and caused grasses to sprout throughout Ovamboland, giving livestock and its owners some respite. In June, however, all cattle was herded far away from the villages, scattered

[34] NAN, NAO 18, Monthly Reports Ovamboland, July–December 1926.
[35] NAN, NAO 18, Monthly Reports Ovamboland August 1927–January 1929.

over a much larger number of cattleposts than previously for fear that the water reserves at the posts again might be insufficient. Testament to the water and grazing scarcity throughout the region was the fact that wildlife concentrated on the banks of the Kunene River, which itself was much reduced in flow. The 1929/1930, 1930/1931, and 1931/1932 rainy seasons were so poor and water reserves at the villages so scarce that many households kept their animals at the cattleposts almost continuously. Only December 1932 brought good rains that allowed the depleted water stores to be recharged allowing the herdsmen to bring the herds home.[36]

In May 1934, toward the end of the rainy season, water and grazing for livestock were abundant and animals remained in good shape throughout the September–October dry season because of the continued availability of water and grazing. The 1934/1935 rainy season ended very early in March 1935 and water was in short supply by the end of April. Good rains in November–December 1935 and February–March 1936 caused the floodwaters from Angola to reach Ovamboland for the first time since the 1928/1929 droughts. Rains and floods filled the seasonal water courses and pans, especially in the Oukwanyama district and the northwestern districts. Pans and flood channels were still full at the end of May 1936 and the Angolan floodwaters had reached the Ekuma in the far south of the floodplain, which fed into Etosha Pan. Grazing was plentiful and livestock was in excellent shape. By August, however, all cattle had been herded back to the remote cattleposts because the open surface water had run out. By the end of October 1936, water and grazing for livestock were scarce "as is usual in this time of year... [and livestock were] consequently in poor condition."[37]

Because of scant rainfall early in the rainy season, herdsmen were cleaning and re-digging the water holes at some of the cattleposts at the end of February 1939 in preparation for an anticipated early trek of the herds to the cattleposts. Such preparations typically were reserved for the June–July months after the harvest. By the end of May, however, no standing water was left in the pans in and near the villages. By September 1939, the shortage of water for livestock caused the colonial administration to initiate relief projects, including digging wells to establish new cattleposts in the northwestern districts. Water and grazing for livestock remained a concern in some of the northwestern districts until the start of the next

[36] NAN, NAO 18–19 f. 11/1, Monthly Reports Ovamboland, February 1929–December 1932.

[37] NAN, NAO 19, Monthly Reports Ovamboland, April/May 1934–October 1936.

rainy season in November–December 1940, when "rains . . . considerably improved the *watering* [of livestock presumably]."[38] The 1940s and early 1950s saw villagers respond in similar highly flexible ways to local rainfall and regional flood events by expanding and maintaining a water harvesting and water storage infrastructure and closely managing the available dry season water stores.[39] The main strategy they used to anticipate water shortages involved temporarily herding one of the major water consumers – cattle – from the villages during most of the dry season (and sometimes longer). The strategy, however, was costly because it deprived villagers of a major source of sustenance and wealth, and, moreover, did not always insulate villagers from drought or high livestock losses.

Woodland Coppice, Clones, and Environmental Infrastructure

Even as towering fruit trees marked many of Ovamboland's farms by the 1960s, colonial experts stressed that the off-farm woody vegetation in and around the villages consisted of bush, a degraded remnant of a past natural forest. Throughout southern Africa, colonial officials identified bush encroachment as a major environmental and economic problem, as was the case, e.g., in South Africa's Kuruman region. Wildlife's decline, livestock's increase, cutting timber trees for use as mine props, and discouraging burning were all factors that combined to transform what appeared to be an open park-like landscape with grass and large trees into bush veld in early nineteenth-century Kuruman. Whereas colonial officials perceived the changes in the Kuruman environment as negative, local African farmers regarded *Acacia mellifera* and other "bush encroachers" as valuable sources of browse for their goats.[40]

Ovamboland's officials held the local populace responsible for the degradation of the forest lands. They were right in that the clearing of woody vegetation was dramatic when refugees and migrants streamed into the middle floodplain in the 1920s and 1930s. But they were also wrong because the environmental transformation that occurred cannot be understood only in terms of linear deforestation. Wood harvesting beyond

38 NAN, NAO 20–21, Monthly Reports Ovamboland, January 1939–July 1941. Emphasis author's.

39 NAN, NAO NAN, NAO 20–21, Monthly Reports Ovamboland, January 1939–July 1941 and Quarterly Reports Ovamboland, January–March 1942 – April–June 1943, NAO 60–61, Quarterly Reports Ovamboland January–March 1944 and January – September 1948, April–June 1949, and April–June 1950 – July–September 1954.

40 Jacobs, *Environment, Power, and Injustice*, pp. 4–5, 97–99, 127–129.

the areas selected as actual fields was less destructive and less permanent. African settlers cut massive amounts of poles and branches, but many woody vegetation species in north-central Namibia had the ability to re-sprout. In fact, the region's ubiquitous bushland – especially its mopane bushland – is coppice woodland maintained through vegetative reproduction. Rehearsing the arguments made in previous chapters on Ovamboland's fruit trees and its dark earths, the human-created coppice woodland constituted neither artifact nor natural resource, and nor unambiguous environmental degradation or improvement. Coppice management is highly selective, privileging certain species over others (and reducing species diversity) and altering the vegetation's architecture (homogenized coppice stools vs. a diversity of bush and tree shapes), deeply transforming the environment, affecting flora and fauna.

The Western seed paradigm that is fixated on and that privileges seed-based sexual reproduction undervalues the importance of asexual vegetative reproduction, even though, ironically, the latter is the more common and more effective method of propagation for many tropical (and non-tropical) plants. The very definition of domesticates has a clear bias in favor of sexual reproduction and evolution-as-progress. Domestic species thus are defined by two characteristics: they are dependent on human assistance for reproduction; and production has been improved (e.g., they yield more or larger fruit) through selection and breeding. Sexual reproduction facilitates human manipulation, control, and improvement.[41]

[41] For examples of the strong emphasis on seed/sexual reproduction, see K. J. Frey (ed.), *Historical Perspectives in Plant Science* (Ames: Iowa State University Press, 1994), e.g. chapter 7 by J. W. Dudley, "Plant Breeding – A Vital Part of Improvement in Crop Yields, Quality and Production Efficiency," pp. 163–177. See also Tomlinson and Zimmermann, *Tropical Trees as Living Systems*, chapters 3–6. Piot and Le Houérou acknowledge the possibility of propagation by cuttings, but they reject it as too costly and too complicated, J. Piot, "Management and Utilization Methods for Ligneous Forages: Natural Stands and Artificial Plantations," and H. N. Le Houérou, "Planting and Management Methods for Browse Trees and Shrubs," in Le Houérou, *Browse in Africa*, p. 345 and pp. 351–359 (especially 353), respectively. In agroforestry projects, the emphasis was on seed nurseries and seedlings. Kerkhof, e.g., hardly refers to vegetative propagation in his overview of agroforestry; see Kerkhof, *Agroforestry in Africa*. In Franzel et al., the overwhelming emphasis is also on seed propagation, see Franzel et al., *Development and Agroforestry*. In the southern African miombo woodlands, most regeneration is through coppice and root suckers, E. Chidumayo and P. Frost, "Population Biology of Miombo Trees," in Campbell, *The Miombo in Transition*, pp. 64–71. Kajembe noted that in Central Tanzania, when local farmers planted trees, they did so mainly through cuttings, Kajembe, *Indigenous Management Systems*, p. 110. Most regeneration in British woodlands was through coppice and suckers, O. Rackham,

Western images of the wild non-West intersect with the seed paradigm. Vegetative propagation marks the fields of horticulture and plantation agriculture, and although some have identified it as cutting-edge modern technology for forestry, non-Western indigenous vegetative propagation – if it is acknowledged at all – is seen to be a hallmark of primitivism.[42] At best, the use of vegetative propagation qualifies the species involved as semidomesticated, implying a preliminary stage of development on a unilinear path to domesticated status within the framework of a Nature (wild)-Culture (domesticated) dichotomy.[43] But, the state of domestication, e.g., may not be an evolutionary high point or dead end: examples of domesticates gone wild abound.[44]

Ovamboland's vegetation is heavily dominated by mopane. Much of the mopane vegetation in the floodplain consists of low bush with occasional tall trees: the low bush is considered "stunted" because it is cut at ground level and subsequently re-sprouts; the tall single stem timber quality trees are considered the natural or climax form of mopane.[45] The occurrence of the tall trees among the bush, however, is not by mere chance. Instead, mopane bush interspersed with the occasional full-grown tree is the typical appearance of "coppice-with-standards"

Trees and Woodland in the British Landscape (London: J.M. Dent, 1993 [1976]), p. 8. Half of the species listed in ICRAF's Multipurpose Trees Database were recorded as potential vegetative reproducers, see P. A. Huxley, "Multipurpose Trees: Biological and Ecological Aspects Relevant to their Selection and Use," F. T. Last (ed.), *Tree Crop Ecosystems* (Amsterdam: Elsevier, 2001), pp. 19–74 (especially 56).

42 J. Westoby, *Introduction to World Forestry*, (Oxford: Basil Blackwell, 1989) p. 13. A number of important plantation trees are vegetatively propagated; e.g., the rubber tree *Herva brasiliensis* and Banana (*Musa* spp.). See Hallé, Oldeman, Tomlinson, *Tropical Trees and Forests*, pp. 25, 121–122, 124–125.

43 Fairhead and Leach, *Reframing Deforestation*, p. 193 reject the usefulness of the domesticate/wild dichotomy. Alcorn rejects the dichotomy and points out that cultivated plants may not always be domesticated and domesticated plants may not always be that carefully cultivated. See J. B. Alcorn, "Huastec Noncrop Resource Management: Implications for Prehistoric Rain Forest Management," *Human Ecology*, 9 (1981), pp. 400–401. Bonnéhin considers the cultivation of wild plants as a second stage of domestication and notes that genetic manipulation of plants (a third stage of domestication) may be too costly for farmers. See Bonnéhin, "Domestication paysanne des arbres fruitiers forestiers," pp. 1–2, 16–17, 61, 67.

44 See Chapter 1.

45 See, e.g., Erkkilä and Siiskonen, *Forestry in Namibia*, pp. 47, 127, 169, 174–177, 179, 183; A. B. Cunningham, "Low-Cost Housing Needs, Wood Use and Woodlands," in G. D. Pierce and D. J. Gumbo (eds.), *The Ecology and Management of Indigenous Forests in Southern Africa* (Proceedings of an International Symposium, Victoria Falls, Zimbabwe, July 27–29, 1992), pp. 283–291.

woodland that results from deliberate human management.[46] Moreover, contrary to being in decline, the mopane vegetation may, in fact, be a more dominant species – especially in the middle floodplain – and it may actually be yielding a higher biomass production now than it did in the 1890s, although at the expense of biodiversity. Rackham has demonstrated that England's woodland cover from the Middle Ages through the early twentieth century shows remarkable continuity despite high levels of wood consumption. One of the principal factors that made this possible was that much of the woodland was heavily managed, principally in the form of coppice-with-standards, making the coppice woodlands a cultural landscape. The regularly coppiced bush was interspersed with occasional tall full-grown trees to provide timber and to facilitate seed production. Although some have argued, e.g., that the enormous wood demand for iron smelting deforested the English environment since the Middle Ages, Rackham identifies coppice management as the technology that enabled England to "sustainably" rely on wood for construction and domestic and industrial fuel consumption.[47] But, the English Medieval coppice woodlands must have been very different in composition and structure from its pre-coppice management predecessors if only because certain species show more coppice vigor than others, humans preferred specific coppice species for their needs, and coppice management produces clones, genetic copies of the parent material: all factors that significantly reduce woodlands' biodiversity. Rackham's sustainability claim is perhaps too strong in that context because the introduction

[46] Heavy elephant browsing also causes a decline of adult trees and a relative increase of coppice. Water logged soils tend to lead to stunted growth of mopane. See Dale. M. Lewis, "Observations of Tree Growth, Woodland Structure and Elephant damage on *Coleospermum mopane* in Luangwa Valley, Zambia," *African Journal of Ecology*, 29 (1991), pp. 207–221.

[47] Rackham, *Trees and Woodland in the British Landscape*, pp. 59–90. See also, M. Jones, "The Rise, Decline, and Extinction of Spring Wood Management in South-West Yorkshire," in C. Watkins (ed.), *European Woods and Forests: Studies in Cultural History* (Wallingford, UK: CAB International, 1998), pp. 55–71. Trees and other plants grow from meristems, cell tissue that can divide. Apical meristems are located in the stem, buds, and root tips, but meristem tissue is distributed throughout the tree between its wood and bark tissue. See Kozlowski, Kramer, Pallardy, *The Physiological Ecology of Woody Plants*, p. 7. Coppice management was also known in China involving oak and *Cunninghamia* (*lanceolata*) and in Japan. For China, see Menzies, *Forest and Land Management in Imperial China*, pp. 95–96, 103, 106. For Japan, see C. Totman, *The Green Archipelago: Forestry in Pre-Industrial Japan* (Athens, OH: Ohio University Press, 1998 [1989]), pp. 36–37, 118–128, 152–153.

of coppice management resulted in the loss of whatever vegetation predated it. It was sustainable only in the sense that the coppice woodland production could be maintained without destroying the coppice vegetation itself. Hundreds of years of coppice management may have resulted, however, in a severe narrowing of the genetic diversity for the species involved, giving rise to a genetic bottleneck that may critically endanger the future of the species, as is the case with the cheetah (equally perhaps a side-effect of human use and management through "domestication" or "taming").[48]

Still, the significance of coppice management of woody vegetation as a practice and a source of tree products has been widely overlooked. For example, critics have argued that the fuel-wood crisis as a contributing factor to deforestation in Africa has been exaggerated because much of the fuel wood is harvested by cutting poles and sticks at ground-level (coppicing) or by cutting branches (pollarding), which allows for regrowth.[49] Many African woody species demonstrate strong coppice regrowth and in the *miombo* woodlands, regeneration through coppice regrowth and root suckers is much more prevalent than seed propagation. Maintaining a few seed-bearing adult plants is important to long-term use because repeatedly harvesting shoots prevents or reduces seed production.[50]

Depending on the length of the coppice, trials with indigenous umbrella thorn (*Acacia tortilis*) in southeastern Botswana produced a coppice regrowth of 1.8 to 9.4 ton per hectare per year between 1976 and 1985. During the same trials, the Australian native *Eucalyptus camaldulensis*, which is also known for its coppice vigor, showed a coppice regrowth of 1.25 tons per hectare per year. But 75 percent of the eucalyptus trees

48 On the cheetah's genetic bottleneck, see Chapter 1.

49 M. Leach and R. Mearns (eds.), *The Lie of the Land: Challenging Received Wisdom on the African Environment* (Portsmouth, NH: Heinemann, 1996), pp. 10–12.

50 G. D. Piearce, "Natural Regeneration of Indigenous Trees: The Key to Successful Management," in Piearce, and Gumbo, *The Ecology and Management of Indigenous Forests in Southern Africa*, pp. 114, 116. On the miombo, see E. N. Chidumayo, "Silvicultural Characteristics and Management of Miombo Woodland," in the same volume, pp. 124–133. See also Chidumayo and Frost, "Population Biology of Miombo Trees," pp. 64–71. Gillet and Piot stress that overexploitation of woody vegetation in the Sahel causes deforestation but they acknowledge that a number of species show strong coppice and lopping regrowth. See H. Gillet, "Observations on the Causes of Devastation of Ligenous Plants in the Sahel and Their Resistance to Destruction" and J. Piot, "Management and Utilization Methods for Ligneous Forages: Natural Stands and Artificial Plantations," in Le Houérou, *Browse in Africa*, pp. 127–129 and p. 344, respectively. Pollarding and coppicing are also used in central Tanzania; see Kajembe, *Indigenous Management Systems*, p. 114.

PHOTO 4. Mopane coppice (photo by the author, 1992).

perished during the trial period, resulting in a loss rate that was seven to ten times higher than that of the acacia woodland coppice.[51]

Bush, and in particular mopane bush, pervades late-nineteenth- and early-twentieth-century descriptions of the Ovambo floodplain. In early twentieth-century Ombadja in the far northwest of the floodplain, ridges with dense bush and dispersed trees separated the wide flood channels. The bush became less dense close to the Ombadja heartlands, but the easily water-logged middle slopes that were less suitable for crops were covered with dense stands of young mopane and other bush, including thorn bush. Moreover, the bushland also contained occasional large mopane and tamboti (*Spirostachys africana*) trees.[52] A missionary

[51] T. Tutema, "Possibilities for the Management of Indigenous Woodlands in Southern Africa," in Piearce, and Gumbo, *The Ecology and Management of Indigenous Forests in Southern Africa*, pp. 134–142. Erkkilä and Siiskonen stressed the enormous potential of (scientific) coppice amangement of mopane; see Erkkilä and Siiskonen, *Forestry in Namibia*, pp. 179 and 183.

[52] Lima, *A Campanha*, pp. 14–18, 111–122, 175–183, 200–203; CNDIH, Avulsos, Caixa 3703, Processo Missão de Estudos no Sul de Angola, Relatório do Mez de Outubro [1914] (this report ascribed the shrub size of woody vegetation to aridity) and Caixa 4130, Governo do Distrito 31–20.1, Relatório sobre a Occupação, Lubango May 22, 1909; Petrus Shanika Hipetwa, interview by the author, Oshiteyatemo (Namibia), June 17, 1993.

traveling through the 1890s northern floodplain Oukwanyama kingdom noted that everything not dedicated to crop fields was covered with bush.[53]

A 1928 description of the vegetation of the middle floodplain, along the Angolan-Namibian border line, offers a virtual transect of the vegetation cover from west to east:

> Between Points [border beacons] 2 and 3 worthless sandveld starts, overgrown with average bush and poor grazing. At point 4 gray loam soil appears in some "laagtes" [low areas, i.e., floodchannels] which soon predominates and continues up to point 6. This part, overgrown with Mopane bush, has also only a little poor grazing.... From Point 6 to 12 the country is nearly always the same – sparse Mopane bush on gray soil, which is often thinly covered with sand and there is to certain extent often very heavy sand in the vast treeless "laagtes" [floodchannels]. There are also a few dune-like parts.... Between Points 14 and 16 one comes across several distinct omurambas [floodchannels] from 300 to 1000 m[eter] broad, between which dunes up to a height of 10 m[eter] lie. The omurambas are devoid of bushes and trees, while the interspersed dunes are thickly overgrown with bush. Mopane is no longer predominant. The first Ukuanyama [Oukwanyama] werft [farm] is at Point 16. The country becomes more open; the Oschanas [floodchannels] have wider but less distinct courses, they are lost in a confusion of branches, islands, "laagtes" and flats. The werfts are on the dune-like banks; the bush has been destroyed. Everywhere one sees the beautiful Onjandi [jackalberry] trees or wild figs.[54]

The challenges of keeping open the cut line that marked the Angolan-Namibian border and Ovamboland's roads attest to the resilience of the vegetation and suggest that bush encroachment with invasive woody vegetation potentially is as much a threat in Ovamboland as it is further south in the Tsumeb and Grootfontein regions. A 1928 report noted that the border cut line was almost indistinguishable at the end of the rainy season and a follow-up report added that keeping the borderline open was difficult because "[t]he scrub, particularly mopane, grows exceptionally quickly."[55] Because of their incessant efforts to keep the

53 AVEM, RMG 2599 C/i 19, Bernsmann, Omburo, January 6, 1892.

54 NAN, KAB 1, Volkmann, October 30, 1928, "Report on the Agricultural and Political Conditions at The Angola Boundary"; RCO 9, RCO to Sec. Protectorate, February 18, 1917; RCO 10, RCO and Hahn, Preliminary memo re. Ovamboland and Chief Mandume, [1916].

55 NAN, NAO 17, UGR to O/C NAO, Namakunde, April 18, 1928, O/C NAO to Sec. SWA, December 24, 1927, to Clarke, Ondangwa, March 4 and April 11, 1928, and to Charlie, Ondangwa, April 11, 1928. See also, NAO 17, NCO to Sec. SWA, Ondangwa, May 2, 1933, and NCO to Clarke, Ondangwa, May 3, 1934, and NCO to Sec. SWA, Ondangwa, July 22, 1937; NAO 18, 20–21, Monthly Reports Ovamboland, September

Angolan-Namibian borderline free of tree and bush vegetation, colonial officials were painfully aware that mopane in particular had remarkable coppice vigor and that it could not be permanently removed unless it was burnt or the stumps dug up and removed.[56] In 1927, a report concluded that what appeared to be the encroachment of mopane bush was contributing to silting up the floodplain's seasonal watercourses and the author of the report suggested clearing the flood channels with dam-scrapers and plows.[57] In fact, a missionary attested that the resilience of mopane was one of the reasons that the Ovambo regarded mopane to be a tree with "spirit." Native Commissioner Hahn in 1935 emphasized that mopane "was one of the most useful trees they [the Ovambos] have, and is used for all kind of purposes."[58] Floodplain farmers principally cut down mopane and tamboti for construction purposes and to clear land for fields. The head of the colonial administration in Ovamboland commented in 1931 that there was "fortunately an abundant supply" of these species.[59]

Woodland Harvesting

Farms on the lower slopes of the ridges between the flood channels backed up and often extended into what in the middle and northern floodplain

1925, June 1939, March–November 1940, January–July 1941, and Quarterly Reports Ovamboland, January–March 1942, April–September 1943, April–June 1946; NAO 25, NCO to Sec. SWA, Ondangwa, December 18, 1942; NAO 43, NCO to Graig, Ondangwa, June 22, 1943; NAO 60–61 Quarterly Reports Ovamboland, April–September 1949, October–December 1951 and April–June 1953.

56 NAN, KAB 1, Submissions to Administrator, Secretary, & Attorney-General of SWA 1927, C.N. Manning (Magistrate Rehoboth), Secretary 1926 and 1927 Commission, Rehoboth, December 15, 1927, to F.P. Courtney-Clarke, Assistant-Secretary SWA, Windhoek. On the coppice vigor of woody vegetation, especially mopane, see also NAO 17, O/C NAO to Sec. SWA, December 24, 1927, and KAB 1, Submissions Manning to Administrator, Secretary, & Attorney-General of SWA, Rehoboth, December 15, 1927; SWAA 3, Administration, Forestry: Indigenous Forests Ovambo A1/2 (I), NCO to CNC, Ondangwa, June 2, 1941. Kaulikalelwa Oshitina Muhonghwo, interview by the author, Ondaanya (Namibia), February 1, 1993.

57 NAN, KAB Volume 1, Draft Report Kunene Water Commission, Olusandja, July [1927].

58 AVEM, RMG 2630 C/k 7, C. Sckär, "Kurze Geschichte der Ovakuanjama," Appendix to C. Sckär to Inspector, Namakunde, October 2, 1912. See also NAN, A450, 9, "Hahn, Rough notes," p. 57. In 1935, Hahn emphasized that mopane "was one of the most useful trees they [the Ovambos] have, and is used for all kind of purposes," A450, 12, SWA Commission: Minutes of Evidence vol. 12, Ukualuthi, August 13, 1935, pp. 649–650.

59 NAN, NAO 44, NCO to Sec. SWA, Ondangwa April 20, 1931.

was known as *omufitu* bush land or forest that covered the crest of the ridges. Settlers cleared farm plots on the lower slopes on both the western and eastern sides of the ridges, which in the floodplain run north-south. Before the 1950s, the *omufitu* bush land on the crest of the ridges had been used as a "natural" boundary between neighboring villages.[60]

The village forests were a main source of wood for constructing new kraals and fences and for securing replacement poles, sticks, and thorn branches.[61] Early-twentieth-century sources emphasize the often very dense bush vegetation on the low ridges between the watercourses.[62] *Omufitu* was the source especially of sickle bush (*Dichrostachys cinerea*), wild seringa (*Burkea africana*), and tamboti, as well as mopane and Transvaal teak (*Pterocarpus angolensis*).[63] Dense stands of mopane bush occurred on locations on the ridges that were easily waterlogged. Thorn bush occurred as well, but thorn trees were relatively rare.[64]

A mopane *omufitu* forest that separated Omduda and Onaihanga villages in early 1941 consisted of tree and bush vegetation. From the eastern edge of the forest, where a young man had been killed in a quarrel, the sparse mopane trees and bushes did not obscure the view of the accused killer's father's homestead, three-quarters of a mile (1,000 meters) away. Visibility in all other directions was poor, although the victim's homestead was half a mile (700 meters) away on the western side of the forest. Bush vegetation between the various trees on the crime site was only two feet (0.6 meters) high. When the crime transpired, a group of young men had been cutting poles to rebuild a homestead that had been moved to a new site within the farm plot. The two-foot high bush may have been

[60] Helemiah Hamutenya, interview by the author, Omuulu Weembaxu (Namibia), July 17, 1993; NAN, NAO 46, Testimony Hamnyela Nashipili, Grootfontein, May 5, 1941, Statements Hamnyela Nashipili, Dirk Jacobus Greyling, Martin Kapenda, Namkoloka Nashipili, Ondangwa, February 12, 1941.

[61] On *omufitu* as the source of replacement poles, see interviews by the author: Moses Kakoto, Okongo (Namibia), February 17, 1993; Nahandjo Hailonga, Onamahoka (Namibia), February 4, 1993; Franscina Herman, Odibo (Namibia), December 12, 1992; Julius Abraham, Olupito (Namibia), June 16 and 18, 1993; NAO 46, Testimony Hamnyela Nashipili, Grootfontein, May 5, 1941, Statements Hamnyela Nashipili, Dirk Jacobus Greyling, Martin Kapenda, Namkoloka Nashipili, Ondangwa, February 12, 1941.

[62] NAN, KAB 1, W. Volkmann, October 30, 1928, "Report on the Agricultural and Political Conditions at the Angola Boundary."

[63] Interviews by the author: Mateus Nangobe, Omupanda (Namibia), May 24, 1993; Malita Kalomo, Omutwewondjaba (Namibia), July 15, 1993; Pauline, Onenghali (Namibia), December 15, 1992; Julius Abraham, Olupito (Namibia), June 16 and 18, 1993.

[64] Petrus Shanika Hipetwa, interview by the author, Oshiteyatemo (Namibia), June 17, 1993.

the regrowth that had sprouted after the poles had been cut. Clearly, the pole cutting had been very selective or the bush vegetation would have been more open.[65]

The colonial government attempted to convert the *omufitu* forests on the low ridges into formal village forest reserves in the 1940s and 1950s, with little or no success. A 1941 letter from the Native Commissioner of Ovamboland to his superior in Windhoek, e.g., stated: "[i]t has always been the policy of this office to encourage the protection of indigenous trees and in every 'umkunda' (area) [village] a portion is, as far as possible, always kept as bush and forest reserve. The difficulty, however, is one of control."[66] In eastern Ovamboland beyond the floodplain, enforcement proved entirely impossible.[67]

But colonial officials' attempts to conserve village forests in the floodplain were similarly thwarted. Pauline and her husband cleared a farm in the *omufitu* forest in the 1940s floodplain village Okatope in Oukwanyama district.[68] In the 1950s, the Oukwanyama district senior headman Gabriel Kautwima permitted farm plots to be cleared in Omhedi village's *omufitu* because little other suitable land for new farms was left. The sandy *omufitu* soil required a large amount of manure before it could be used for grain cultivation.[69] Joshua Mutilifa was the first to settle in the *omufitu* forest of Omhedi, but many followed his example and by the 1990s the forest virtually had disappeared.[70] This story repeated itself in other villages: by the early 1990s, the Oukwanyama village of Oshomukwiyu, south of Omhedi and settled at a later date, also lacked any *omufitu* forest. In fact, very little off-farm bush remained and even the stumps of the previously abundant mopane vegetation had been dug out for use as firewood, thereby preventing the regeneration of the hardy plant. Kulaumoni Haifeke, who grew up in Oshomukwiyu, recalled that the area had once resembled uninhabited "wilderness," and explained that people had cut all the trees and bushes in "no-man's land" (i.e., in the village "commons"): "[y]ou can't prevent it because they are not on your farm. People cut them in places which are not owned." Asked

65 NAN, NAO 46, Testimony Hamnyela Nashipili, Grootfontein, May 5, 1941, Statements Hamnyela Nashipili, Dirk Jacobus Greyling, Martin Kapenda, Namkoloka Nashipili, Ondangwa, February 12, 1941.

66 NAN, SWAA 3, Administration, Forestry: Indigenous Forests Ovambo A1/2 (I), NCO to CNC, Ondangwa, June 2, 1941.

67 NAN, NAO 44, ANC to NCO, Oshikango, March 24, 1942.

68 Pauline, interview by the author, Onenghali, (Namibia) December 15, 1992.

69 Interviews by the author: Joshua Mutilifa, Omhedi (Namibia), March 8, 1993, and Nahandjo Hailonga, Onamahoka, February 4, 1993.

70 Joshua Mutilifa, interview by the author (Namibia), March 8, 1993.

whether rules had prevented such behavior in the past, she responded: "[t]here was no such rule to prevent people to cut down trees outside anybody's land."[71]

At the same time that the clearing of new farms literally cut into the village forest lands, on-farm arable land was expanded at the expense of the on-farm woodland and bush fallow, as a result of either subdividing farms or enlarging the fields. When Nahango Hailonga separated in the 1940s, e.g., her ex-husband and their sons cleared a farm plot for her in his bush fallow. Even in the expanses of eastern Ovamboland, where villages typically were isolated from one another by large stretches of bush and forest land, fields encroached on farm fallow. In the 1950s, Joseph Kambangula extended his fields into the on-farm bush fallow that had been maintained by his late father. In the 1960s, after purchasing an existing farm, Moses Kakoto cleared and plowed the fallow that the previous owner had used as a source of forage. The woody vegetation in the bush fallow supplied critical resources; shrinking fallow diminished the availability of on-farm construction wood, fuel, and forage.[72]

The newly appointed agricultural officer for Ovamboland in 1957 not only considered the bush vegetation in and around the villages to be useless but also identified the species it contained – in particular "dwarf" mopane bush, black thorn (*Acacia mellifera*), and buffalo thorn (*Ziziphus mucronata*) – as dangerous bush encroachers. He noted that even in densely inhabited areas where other species had become rare, these "undesirable" species appeared abundant. Apparently unaware of previous attempts to create village forest reserves, he erroneously ascribed their abundance to their unsuitability as construction materials and fuel, even though the opposite was true.[73] Another contemporary report emphasized that although mopane was cut in large quantities, it was known to regrow fast.[74] The South African Deputy Secretary for Forestry visited Ovamboland in 1961 and, traveling along the border road from east to west (i.e., from eastern Ovamboland toward the floodplain), observed

[71] Kulaumoni Haifeke, interview by the author, Oshomukwiyu (Namibia), May 11, 1993.

[72] Interviews by the author: Joseph Kambangula, Omboloka (Namibia), February 25, 1993; Nahango Hailonga, Onamahoka (Namibia), February 4, 1993; Timotheus Nakale, Big Ekoka ((Namibia), February 21, 1993; Moses Kakoto, Okongo (Namibia), February 17, 1993. BAC 133, Quarterly Report Agriculture Ovamboland, June 30, 1958.

[73] NAN, BAC 131, Agricultural Officer Ovamboland to Bantu Commissioners Ondangwa and Oshikango, January 28, 1957.

[74] NAN, SWAA 3, Acting Sec. SWA, "Preservation of trees," May 14, 1957.

that twenty to twenty-five miles east of Oshikango (the far edges of the floodplain), the vegetation changed radically from a "more or less closed forest" east of that point to a much more open landscape with a few large fruit trees dotting fields and stressed that "the landscape is really all grassland with large patches of factually pure mopaniveld in the form of bush."[75] The grasslands were in the flood channels and the fields, and the bush marked the higher ground where the villages were located. In 1970, during a brief consultancy to the area, a South African expert noted that the large mopane bushlands between Oshakati and Ombalantu "appear undeveloped and effectively useless, except perhaps as emergency pasturage," although in fact they must have been very heavily used over a sustained period.[76]

To ensure a supply of mopane wood, households relied on its coppice vigor; in addition to soil and other conditions, this factor would also explain why mopane stands typically consisted mostly of brushwood and small trees with only occasional larger trees of seed bearing age.[77] The forestry department's reports about research conducted on mopane coppice vigor in the 1970s provide some details about mopane bush land in the region. In 1976, five small demarcated plots that included patchy spots that were separated by farms near Onankali, fifty-five miles south of Ondangwa on the road to Tsumeb, measured 10,600 square meters combined, which is a little more than a hectare. The area contained 2,743 mopane bushes with an average height of six to eight feet (1.8 to 2.3 meters). Assuming that there was one suitable pole per stool (and mopane stools typically have multiple stems), the Onakali mopane bush therefore contained 2,588 suitable construction poles per hectare. After the measurements had been taken, the mopane bush was subjected to different regimes of coppice and thinning management. The trials indicated that with careful management, mopane could produce good small poles for use in wire fences (the palisade poles were thicker and longer than these) in three to four years.[78] The palisade poles were approximately

[75] NAN, BAC 131, Deputy Secretary of Forestry, "Report of a visit by the Deputy Secretary of Forestry to South West Africa: April 17–29, 1961," Pretoria, May 10, 1961.

[76] NAN, OVA 57, Report on visit to SWA, November 3–15, 1969, Appendix to Dr. H. A. Lueckhoff to Chief Director Department of Bantu Affairs, Ondangwa, April 3, 1970.

[77] The reliance on coppice rather than seed is suggested by Erastus Shilongo, interview by the author, Olupanda (Okalongo, Namibia), June 21, 1993.

[78] NAN, OVA 58, Department of Agriculture and Forestry to Department of Forestry Pretoria, Ondangwa, September 14, 1976, and Department of Forestry Pretoria to Secretary of Agriculture, Owambo, Pretoria, June 24, 1976.

two to three times thicker than and twice as long as the fence poles and would have taken longer to grow.

At Ogongo, four smaller mopane trial plots were established and measured within a larger plot that had been laid out in 1971 at the site of the Agricultural College. Measurements showed that per hectare, the plot contained on average 174 mopane trees, 865 mopane bushes, and 608 mopane saplings. Presuming the trees and bushes yielded a palisade-size construction pole each, the Ogongo mopane bush produced approximately 1,000 poles per ha.[79]

The experiments were short-lived: the plots at Onankali were cut clear in August 1976, but not before mopane coppice vigor had impressed the forestry officials.[80] Both the Onankali and Ogongo mopane trial plots consisted of patches that were (in the case of the former) and had been (in the case of the latter) close to farms, and they therefore would have been heavily used as a source of construction and firewood before the plots had been appropriated for colonial experiments.[81]

[79] NAN, OVA, D.P.J. Opperman and C.L. Prinsloo, Botanical Survey and Physical Planning of the Orongo [Ogongo] Trial Area, Ovamboland, Appendix to D. Opperman to Director Agriculture Ondangwa, Windhoek, July 28, 1969. The density per hactare was derived from the number of plants counted in 25 circle quadrants with a radius of 30 feet each. The total surface of 25 circle quadrants with a 30-foot radius is 0.785 ha. The figures provided in the report were thus multiplied by a factor of 1.27. A survey in mopane-dominated vegetation in a forest reserve in Malawi consisted of 18% trees, 47% shrub, and 35% herbs, and counted 556 trees/ha. The vegetation was exposed to grazing and harvesting of poles and firewood and charcoal. Mopane also occurred outside of the reserve but was under heavy pressure as a result of agricultural clearing. See Chikuni, "Conservation Status of Mopane Woodlands in Malawi," pp. 250–258.

[80] NAN, OVA 58, Agriculture and Forestry Ondangwa to Forestry Pretoria, September 14, 1976 and Sec. Dept. Forestry Pretoria to Sec. Agriculture and Forestry Owambo, June 24, 1976; OVA 6, weekly reports, July 9, 16, and 30, and August 27, 1976.

[81] To establish Ogongo Agricultural College, where the mopane plots were located, almost fifty farms were destroyed, NAN, BAC 132, Trust Farming Projects (1960–1962), Agriculture Officer to Bantu Commissioner, Grootfontein, December 22, 1962, Report on an Inspection Tour to Ovambo, December 11–18, 1961; AHE (BAC) 1/346, Report of the SWA Planning Committee for Agricultural Training Centers, Appendix to Chief Bantu Affairs Commissioner SWA to Bantu Affairs Commissioners Ondangwa, Runtu, and Oshikango [Windhoek], April 8, 1965; Bantu Affairs Commissioner to Chief Bantu Commissioner, Ondangwa, December 30, 1965, and Chief Bantu Commissioner SWA to Secretary Bantu Administration and Development, Windhoek, January 11, 1966; and Bantu Affairs Commissioner to Chief Bantu Commissioner, Ondangwa, May 16, 1966. The Onankali mopane plots were separated by Ovambo farms, NAN, OVA 58, Department of Agriculture and Forestry to Department of Forestry Pretoria, Ondangwa, September 14, 1976, and Department of Forestry Pretoria to Secretary of Agriculture, Owambo, Pretoria, June 24, 1976. Erkkilä and Siiskonen stress that mopane is the most popular tree species for construction. They seem to suggest that mopane was

If the trees and bushes are counted as suitable for use as palisade poles, the Ogongo mopane bush would have yielded approximately 1,000 poles per ha, and the Onankali mopane bush more than double that amount at 2,500 poles per ha. If a yield of 1,000 mopane poles per hectare is taken as being more representative of mopane bush land composition in Ovamboland, the construction of the new 6,000 homesteads that refugees from the north founded in the middle floodplain between 1915 and 1933 would have deforested 6,000 ha (in mopane equivalencies), or an average of 352 ha per year to produce the poles for the palisades alone. Constructing the palisades of the 18,386 homesteads existing in 1933 Ovamboland as a whole, with an average 100-meter palisade requiring 1,000 ten centimeters (four inch) thick and 1.8–2.4 meters (6–8 feet) long poles, would have consumed roughly 18,000 ha of mopane bush land. The total surface area of Ovamboland was 4.5 million ha, although only a portion of the area was mopane bush.[82] Assuming that homesteads continued to be constructed in the same way, the 90,918 "traditional" homesteads that were counted in the 1991 Ovamboland census would have required more than four times as much mopane (i.e., 90,000 ha of the Ogongo type mopane bush land or 36,000 ha of the higher-density Onankali type mopane bush land).[83]

In reality, the figures probably overestimate the impact on mopane because many trees other than mopane were used to construct palisades. Still, the 1993 OMITI survey indicated that mopane was by far the most important source of construction wood. Most respondents mentioned mopane as the principal source of wood for enclosures, huts, and fences. Silver cluster leaf (*Terminalia sericea*) was also important, while leadwood (*Combretum imberbe*), tamboti, Rhodesian teak (*Baikiaea plurijuga*), Rhodesian bushwillow (*Combretum mechowianum*), and wild seringa (*Burkea africana*) were minor sources of poles.[84]

mainly used in the form of adult full-grown trees, but coppice poles served as the principal construction wood. See Erkkilä and Siiskonen, *Forestry in Namibia*, pp. 151–152. Erkkilä's 2001 study also appears to highlight the use of larger and smaller mopane (and other) *trees*, rather than coppice bush. See Erkkilä, "Living on the Land," p. 52.

82 NAN, OVA 43, [Dr. J. G. V. Joubert], Final Report "Research on the Potential and Use of Shana pastures in Owambo," Appendix to Sec. Plural Relations [formerly Bantu Affairs] to Sec. Agriculture Owambo, Pretoria, June 5, 1979.

83 Namibia, 1991 Census, Report A, Statistical Tables, vol. v, tables H01 and H02.

84 OMITI 4.3.3, 4.3.9, and 4.3.12. On the importance of mopane as a construction material, see also NAN, BAC 133, Agricultural Officer Ovamboland, Agricultural Report Ovamboland 1955/1956; Erkkilä and Siiskonen, *Forestry in Namibia*, p. 161; Erkkilä, "Living on the Land," p. 100.

The durability of wood types varied. Depending on the species, palisade and other poles needed to be replaced every three to ten years.[85] Colonial reports of the mid-1950s estimated that a homestead's poles were replaced every two to five years, but noted that it had become less common to do so, and that in areas where trees were scarce, grain stalks were used instead of poles. The durability of the most important sources of wood, however, was higher than the colonial estimates. In addition, when a homestead or cattle kraal was moved to a new location on the farm plot, many of the poles were recycled. In 1993, Julius Abraham still had a few of the wild seringa and tamboti poles in his Olupito farm's palisade that his father had cut when he first built his farm on the plot in the late 1920s.[86] Moreover, with careful tree and bush management, not every thornbush branch or pole that was cut was a bush or a tree lost. Where the woody vegetation was not considered to pose an obstacle to growing crops, it was coppiced or pruned with the intention of allowing it to regrow. Kalikalelwa Oshitina Muhonghwo spoke in terms of "cultivating" the bush, an activity that included some light pruning to ensure that the remainder grew well. She "cultivated" mopane, Rhodesian bushwillow, and silver cluster leaf although the silver cluster leaf grew slowly compared to the first two.[87] In the 1993 OMITI survey, tree coppicing was practiced by almost one out of every five respondents' households and pruning was practiced by almost half.[88] If they were cut at ground level, large trees did not survive, with the exception of Rhodesian bushwillow, corkwood (*Commiphora* spp.), and marula, which developed shoots from the roots that were left after their trunks had been removed. Young and small trees that had been cut at ground level recovered slowly, but bushes rapidly regained their former height. This was especially true for jackalberry (Kalahari), apple-leaf (*Lonchocarpus capassa* and/or

[85] For details, see Chapter 3.

[86] NAN, A450, vol. 9, C. H. L. Hahn, "Preliminary Notes on Certain Customs of the Ovambo," *Journal of the SWA Scientific Society*, 3 (1927/28), pp. 16–18; BAC 133, Agricultural Report Ovamboland 1955/1956 and Agricultural Officer to NCO, "Report of travel to the northwestern part of Ovamboland from 20–22 June 1956," Ondangwa, July 4, 1956, and Agricultural Officer Ovamboland to Bantu Commissioners Ondangwa and Oshikango, January 28, 1957; Julius Abraham, interview by the author, Olupito (Namibia), June 16 and 18, 1993. Moses Kakoto used a few new poles when he moved his homestead in 1992, interview by the author, Okongo (Namibia), February 17, 1993.

[87] Kalikalelwa Oshitina Muhonghwo, interview by the author, Ondaanya (Namibia), February 1, 1993.

[88] OMITI 4.4.19 and 4.4.21.

L. nelsii), red bushwillow (*Combretum apiculatum*), leadwood, and Russet bushwillow. Most trees (including mopane) developed shoots when the trunk was cut at sufficient height. When fire was used, however, a tree would not recover. Timing was critical; to ensure regrowth, trees could only be cut down during the rainy season, when they had leaves. One interviewee noted that the older generations possessed this intricate indigenous knowledge about coppice and pruning management, but that it was being lost.[89]

Mopane was among the most frequently coppiced species.[90] When multiple-stem mopane coppice stools were reduced to one shoot to allow a two to four meter (6–12 feet) distance between individual plants, proper ground clearing and pruning meant that mopane shoots had the potential to grow into straight poles of a suitable size for building palisades, huts, and fences within three to five years. In fact, coppiced mopane may produce a higher biomass than uncut mopane. The agricultural officer for Ovamboland noted in 1956 "that where the Mopane is thinned a lot, much Mopane coppice grows."[91] Coppiced silver leaf cluster took two to four years to grow to sufficient size for constructing the palisade (i.e., a length of five to seven feet [1.5–2 meters] and a thickness of about four inches [10 centimeters]).[92] Mopane did not occur east of the floodplain beyond Omutwewondjaba, which may be related to the occasional occurrence of frost, but other tree species (e.g., silver leaf cluster) were employed for similar purposes.[93]

89 Erastus Shilongo, interview by the author, Olupanda (Namibia), June 21, 1993. Zimbabwean research noted that most miombo species coppice optimally if cut higher up the trunk rather than lower down. Chidumayo and Frost consider approximately 1.3 meter (approximately four feet) above ground level best. The research probably refers to coppice from the trunks of adult trees, rather than coppice from a stool; see Chidumayo and Frost, "Population Biology of Miombo Trees," pp. 64–71, and P. T. Mushove and J. T. Makoni, "Coppicing Ability of *Colophospermum mopane*," in Piearce and Gumbo (eds.), *The Ecology and Management of Indigenous Forests in Southern Africa*, pp. 226–230.

90 OMITI 4.4.20.

91 NAN, BAC 133, Agricultural Officer to NCO, "Report of travel to the northwestern part of Ovamboland from June 20–22, 1956," Ondangwa, July 4, 1956. On mopane cutting and higher biomass production, see Mushove and Makoni, "Coppicing Ability of *Colophospermum mopane*," pp. 226–230. Cf. on the three times higher woody plant density in young regrowth miombo compared to old growth miombo, see Chidumayo, "Silvicultural Characteristics and Management of Miombo Woodland," pp. 124–133.

92 Kalolina Naholo, interview by the author, Ohamwaala (Namibia), January 26–27, 1993.

93 Mopane occurs abundantly as far east as Omutwewondjaba but is absent further east at, e.g., Omundaunghilo, personal observations with the author, Omutwewondjaba, July 15, 1993. On mopane brush and its vulnerability to frost that occurs in the east, see,

Pruning trees and bush was practiced by almost half of the households that participated in the 1993 OMITI survey.[94] The species most frequently subjected to pruning were mopane (by far the principal managed species), birdplum, real fan palm, and marula.[95]

In brief, the woodland vegetation in and around the villages was heavily coppiced and pruned, with coppice and pruning cycles that sustained high levels of production of mopane and other species suitable for use in building palisades, huts, and fences. As is the case with north-central Namibia's mopane woodland, southern Africa's ubiquitous mopane woodland may not be degraded woodland, but environmental infrastructure, created and carefully maintained through a coppice regime.[96] Highland Madagascar *tapia* woodlands consisting of scrubby forest reminiscent of southern Africa's miombo are typically also considered to be degraded remnants of a primordial forest that covered much of Madagascar. But Madagascar's *tapia* woodlands are the product of continuous human management. Managed with fire and through coppice, the *tapia* woodlands are the main source of food for Madagascar's wild silkworm and provide fruits, medicine, and firewood.[97] Thus, dense and low "bushlands" may constitute key environmental infrastructure, although they are typically labeled as degraded or useless, and consequently receive little or no attention. For example, South Africa's pathbreaking Nylsvley savanna research project explicitly excluded "tropical thickets" with dense low multi-stemmed growth from the study, considering them unimportant and the mere product of particular poor environmental conditions.[98]

Conclusion

Neither the villages' water nor their woodland resources were "natural"; in fact, all the dry season sources of water and the woodlands were, to a significant extent, not only shaped by human use but were also a human

e.g., Helemiah Hamutenya, interview by the author, Omuulu Weembaxu (Namibia), July 17, 1993. In the surveyed farms in the western part of eastern Ovamboland, Erkkilä noted that only in the far west Ondobe area mopane was used extensively; see Erkkilä, "Living on the Land," p. 86.

94 OMITI 4.4.19 and 4.4.21.

95 OMITI 4.4.22.

96 Mopane occurs as far north as Malawi; see Chikuni, "Conservation Status of Mopane Woodlands in Malawi," pp. 250–258.

97 On Madagascar's *tapia* woodland, see Kull, *Isle of Fire*, pp. 116–144.

98 Scholes and Walker, *An African Savanna*, p. 9.

product. They were not entirely artificial either, in the sense that humans had fully harnessed the villages' water and forest resources. The main source of water for the majority of north-central Namibia's population – water holes – had limited yields and the region's dry season stores of water had to be judiciously managed and strictly rationed. The intense demand on woodland resources could be sustained only through coppice management and a reliance on asexual vegetative reproduction. Human control over sexual reproduction of plants and animals conventionally defines domestication – that is, the control of Culture over Nature.

As was the case at the farm- and field-level of infrastructure, the creation of village-level water and woodland environmental infrastructure was not a one-time investment or the stable outcome of a single process. Water and woodland infrastructure similarly had to be continuously remade and reconceptualized. Water holes, wells, and reservoirs had to be re-dug, repaired, and maintained through annual labor-intensive investments that relied on social processes to mobilize labor through kinship, patron-client, and village networks, creating complicated webs of overlapping rights and obligations. In the case of wells and water reservoirs, individual households simply lacked the capital, labor, and expertise to construct and maintain them on their own. Village forests could only be maintained through continuous coppice management. Thus, north-central Namibia's savanna woodlands, which are generally considered to be wild and stunted bush land rife with dangerous invasive species, are neither natural nor wild. To the contrary, coppice woodland was and is socially reproduced; its form and composition have been shaped by humans who exploited it as a source of forage, fuel, and construction materials. Its very *biological* reproduction was a social process exploiting the species' natural vegetative reproduction potential. In this respect, the southern African bushlands are very similar to premodern European forests, slash and burn forests throughout Africa, Asia, and Latin America, and the infamous *Imperata* grasslands of Southeast Asia, long decried as "green deserts." In fact, they are the result of intensive human management and use and are highly valued as pastures.[99] Bamboo and rattans

[99] On "green deserts," see Goudie, *The Human Impact on the Natural Environment*, p. 37. On the reevaluation of Southeast Asia's long maligned grasslands, see M. R. Dove, *Southeast Asian Grasslands: Understanding a Vernacular Landscape: Canonical Readings* (New York: The New York Botanical Garden Press, 2008). See also D. G. Sherman with the assistance of H. Bruyns, *Rice, Rupees, and Ritual: Economy and Society among the Samosir Batak of Sumatra* (Stanford, CA: Stanford University Press, 1990), pp. 142–159. On the creation of pastures in Southeast Asia, see P. Boomgaard,

PHOTO 5. Recycled poles: Julius Abraham (on the left) with a row of palisade poles originally cut by his father in the 1920s (photo by the author, 1993).

in Asia constitute other intriguing examples. They are considered to be widely used and critical forest products and wild and natural resources, yet their preferred habitat is (human-made) secondary forest, and they are also often cultivated.[100]

"Horses, Horse Trading and Royal Courts in Indonesian History, 1500–1900," and "The Age of the Buffalo and the Dawn of the Cattle Era in Indonesia, 1500–1850," in P. Boomgaard and D. Henley (eds.), *Smallholders and Stockbreeders*, pp. 211–232 and 257–282, respectively. On premodern European forests as human-managed and used, see Watkins (ed.), *European Woods and Forests*.

100 As is the case with mopane, bamboo and rattan are mainly vegetative propagators. On bamboo, see F. Maoyi and Y. Xiaosheng, "*Moso* Bamboo (*Phyllostachys Heterocycla* var. *Pubescent*) Production and Marketing in Anji County, China" and A. Van Bay, "Bamboo (*Neohouzeaua dullooa*) Production and Trade in Cho Don, Vietnam: NTFP Extraction from Allocated Forest Lands," in K. Kusters and B. Belcher (eds.), *Forest Products, Livelihoods and Conservation*, pp. 247–264 and 335–346, respectively. On rattan, see V. D. Quang, "Domestication of Rattan (*Calamus tetradactylus*) in the Buffer Zone of Ke Go Natural Reserve Area, Cam Xuyen District, Vietnam," H. G. Palis, "Rattan (*Calamus* spp.) Extraction in the Philippines: The Case of Manggapin and Kalakwasan Watersheds, Palawan," and F. Pambudhi et al., "Rattan (*Calamus* spp.) Gardens of Kalimantan: Resilience and Evolution in a Managed Non-Timber Forest Product System," in K. Kusters and B. Belcher (eds.), *Forest Products, Livelihoods and Conservation*, pp. 283–293, 313–323, and 347–364, respectively.

The data suggest that other vegetation types that at times have been too easily discarded as degraded and useless deserve more careful study as well, including the secondary (forest) growth that results from shifting cultivation and slash and burn regimes, and burnt and browsed landscapes in general.[101] But, although, in terms of productivity, such environmental infrastructure may be exploited in a "sustainable" way because of heavy and sustained human and nonhuman investments, *environing* is both the cause and the dynamic outcome of significant and often ambiguous and plural environmental changes, both positive and negative, and including but not limited to a substantial reduction of bio- and genetic diversity.

101 Goudie stresses that the quality of secondary as compared to primary forest may be low but it is far from "useless bush." See Goudie, *The Human Impact on the Natural Environment*, pp. 38–39.

5

Browsing and Burning Regimes

Bushland Savanna as Environmental Infrastructure

North-central Namibia's pastoralism heavily shaped the environment. Herders guided their animals through the landscape and wielded the firebrand to burn off old, dry vegetation that they judged to have lost its nutritional value (for their animals) as well as to keep predators at bay. Grazers and browsers trimmed grasses, herbs, and woody vegetation, encouraging the regrowth of young shoots during the rainy season.[1] Animal grazing and browsing also shaped fire regimes by limiting the buildup of fuel: fuel load is critically correlated to the temperature and intensity of fires. The effects of browsing and burning on past and present landscapes are closely related. For example, although cattle grazing in North America appears to have no great impact on species diversity, and burning in and by itself reduces biodiversity, herbivory and burning in combination may increase biodiversity.[2] In many environments, regular burning is a critical tool to create and maintain the grasslands that support livestock. Madagascar's highland grasslands, e.g., are essential in sustaining its large cattle herds, which are still an important source of income, and which were a major export in both the precolonial and colonial eras. Fire was critical to preventing bush encroaching on the grasslands. During the sixteenth through nineteenth centuries, burning and browsing played a key role in the creation and maintenance of extensive *alang alang* grasslands to support large numbers of water buffalo that were indispensable tools in irrigated rice cultivation throughout Southeast

[1] Kull, *Isle of Fire*, pp. 81–115.

[2] On the correlation between fuel load and fire intensity and the impact of combined herbivory and burning, see Whelan, *The Ecology of Fire*, pp. 31 and 285–288, respectively.

Asia.[3] Reducing grass growth by grazing and burning also may favor certain tree species that do not otherwise compete effectively with grasses, including many species identified by the World Agroforestry Centre (ICRAF) as valuable multipurpose trees.[4] Davies emphasizes that the western Mediterranean bush of North Africa is not a degraded relict but constitutes highly resilient vegetation that is well adapted to livestock foraging and fire. Jacobs makes a similar argument about the thornveld bush in the Kuruman region of South Africa which local farmers appreciate as a critical source of smallstock browse.[5] The banning of fire and the exclusion of livestock may have been important factors in bush encroachment on pasturelands in Tanzania's Arusha National Park in the 1980s.[6]

Burning, grazing, and browsing are species-specific, favoring certain plants over others, and locale-specific: within the villages of Ovamboland, e.g., fields were off-limits to livestock for the duration of the cropping season. Small- and largestock grazed and browsed in and near the villages during the rainy season; during the dry season herdsmen took the cattle to remote cattleposts via the cattle trails. The impact of cattle on the environment between the location of the home villages and the sites of the cattleposts was confined to rather narrow corridors. The routes between the villages and the cattleposts were dictated by the availability of water early during the dry season (when cattle was herded to the cattlepost) and early during the rainy season (when cattle was returned to the villages). The spatial and temporal patterns of movement are reflected in when, where, and how fire affected the environment. Fire was not a random or "natural" factor in north-central Namibia, even

[3] Kull, *Isle of Fire*, p. 85–88. On Sumatra, see Boomgaard, "Horses, Horse Trading and Royal Courts in Indonesian History, 1500–1900," and "The Age of the Buffalo and the Dawn of the Cattle Era in Indonesia, 1500–1850," pp. 211–232 and pp. 257–282, respectively. *Alang alang* refers to cogon grass (*Imperata cylindrica*).

[4] Huxley, "Multipurpose Trees," pp. 19–74 (especially 58). On livestock grazing and fuel loads, see C. Toulmin, "Tracking through Drought: Options for Destocking and Restocking," in I. Scoones (ed.), *Living with Uncertainty: New Directions in Pastoral Development in Africa* (London: Intermediate Technology Publications, 1994), pp. 95–115. But fire regulation was not the only factor in bush encroachment in Ethiopia; see D. Layne Coppock, "Vegetation and Pastoral Dynamics in the Southern Ethiopian Rangelands: Implications for Theory and Management," in R. H. Behnke Jr., I. Scoones, and C. Kerven (eds.), *Range Ecology at Disequilibrium: New Models of Natural Variability and Pastoral Adaptation in African Savannas* (London: ODI, 1993), pp. 42–61 (especially 55).

[5] D. K. Davis, *Resurrecting the Granary of Rome: Environmental History and French Colonial Expansion in North Africa* (Athens: Ohio University Press, 2007), p. 15, and Jacobs, *Environment, Power, and Injustice*, pp. 4–5, 97–99, 127–129, and 209–210.

[6] Neumann, *Imposing Wilderness*, pp. 159–160.

as the area's semiarid climate and its vegetation "naturally" created the ideal conditions for seasonal burning: hot and dry with abundant fuel. Instead, fire occurred within the framework of an indigenous burning regime that was grafted onto local environmental conditions. Fire was a principal tool to clear bushland to make fields and a farm. In addition, people burned the crop residue and weed growth on their fields prior to the onset of the rainy season to prepare the land for the next cropping season, thereby contributing to the formation of rich organic dark earth.

The main architects of the browse and burn regime were the animals' owners, users, and managers, even if their decisions were influenced by environmental as well as political, social, economic, and social considerations.[7] But the animals were powerful agents in their own right. The livestock's owners and caretakers decided where the animals would move, how long they would remain in a locale, and whether or not to burn; but the animals did the actual browsing and grazing. The extent of the impact of livestock on the environment is a topic of intense debate and "managed" landscape burning is highly controversial. A declinist model of environmental change emphasizes the negative environmental impact of livestock, highlighting the effects of soil compaction through trampling, overgrazing, deforestation, and desertification. An inclinist literature has emphasized the positive interaction between livestock and vegetation.

Historically, fire was a common environmental management tool and slash and burn methods continue to be used in many parts of the world. Scientific environmental management (especially under the influence of professional forestry) condemned the use of fire as wasteful and destructive and propagated and (brutally) enforced the eradication of the use of fire.[8] In recent years, there has been a reevaluation of the use of fire in conservation, but it has proven difficult to perform controlled burns. Wildfires continue to occur in the American Southwest (despite the United States having superior equipment and firefighting infrastructure), as well as in southern Europe and in southeastern Australia.[9] Because fire has resisted human attempts at domestication, fire conceptually lies in the middle ground between Nature and Culture. A leading historian of fire calls it a "feral force" which "wild or tame... persisted. Humans could neither wholly control it nor live without it." Thus fire is both a

[7] Goudsblom emphasizes that other animals use tools and language but only humans managed to control fire; see Goudsblom, *Fire and Civilization*, p. 1.

[8] Pyne, *World Fire*, pp. 188–212; Pyne, *Vestal Fire*, pp. 106–209; Kull, *Isle of Fire*, pp. 45–76; Scholes and Walker, *An African Savanna*, pp. 111–125.

[9] Pyne, *World Fire*, p. 37.

natural phenomenon and a technology, and is both "metaorganic" and "metacultural."[10]

Livestock and Browse

In dry environments, livestock and wildlife are critically dependent on woody vegetation, and grazing and browsing profoundly shape the vegetation. But the relationship between livestock and the environment in Africa, as elsewhere, has historically been cast in almost exclusively negative terms: overstocking causes overgrazing, which in turn leads to degradation, deforestation, and desertification.[11] Namibian colonial foresters were no exception in warning that livestock numbers in Ovamboland far exceeded the carrying capacity of the land. They argued that livestock trampling under shade trees prevented tree regeneration and lamented "[that] [l]arge parts of Owambo [Ovamboland] are already without foddertrees – the cattle and goats eat all the young trees."[12]

The New Rangeland Ecology literature, however, rejects the declinist overgrazing assessment. It argues that the carrying capacity approach – on the basis of calculating how many livestock units a given environment can sustain – does not apply to arid and semiarid environments because it is based on an equilibrium model. The presumption of the carrying capacity idea is that livestock numbers are the main constraint on the available

[10] Pyne, *Vestal Fire*, pp. 4, 49. Pyne notes that the United States has the best financed firefighting organization of the world, but it cannot prevent wildfires; see Pyne, *World Fire*, p. 37.

[11] H. N. Le Houérou, *The Grazing Land Ecosystems of the African Sahel* (Berlin: Springer-Verlag, 1989), pp. 90–128; H. Gillet, "Observations on the Causes of Devastation of Ligenous Plants in the Sahel and Their Resistance to Destruction," in H. N. Le Houérou (ed.), *Browse in Africa*, pp. 127–129; J. Westoby, *Introduction to World Forestry* (Oxford: Basil Blackwell, 1989), pp. 172–173. Thirgood identified especially nomadic smallstock herding as the cause for a massive and long history of degradation in the Mediterranean; see J. V. Thirgood, *Cyprus: A Chronicle of Its Forests, Land, and People* (Vancouver: University of British Columbia Press, 1987). Davis argues that French colonial narratives about deforesting pastoral nomads developed in North Africa and had a deep impact in France and on its colonies beyond the French colonial empire, influencing especially environmental narratives about the Mediterranean. Including the work of Thirgood, see Davis, *Resurrecting the Granary of Rome*, pp. 134–186, especially 170–174. On the creation of overstocking and environmental degradation narratives, see Anderson, *Eroding the Commons*.

[12] NAN, OVA 57, Dr. H.A. Lueckhoff, "Report on a visit to South West Africa, November 3–15, 1969," Appendix Regional Forester to Director-in-Chief Department of Bantu Administration and Development Pretoria, Grootfontein, April 3, 1970, and Le Roux to Sec. Agriculture, Inventory: Indigenous Forests, Ondangwa, November 5, 1976.

vegetation in an area: too many livestock cause vegetation degradation. The New Rangeland Ecology counters that rainfall is the main constraint on vegetation, and that in environments with high intra- and inter-annual rainfall variation, the system is always in disequilibrium. The concept of carrying capacity under such circumstances becomes meaningless and may convey a false sense of security to pastoralists who heed the carrying capacity recommendations. Moreover, the carrying capacity model presumes that a given quantity of livestock interacts with a given environment, but, many pastoralists in arid and semiarid regions move their livestock, sometimes across large distances, exploiting multiple environments. In fact, under disequilibrium conditions, it is rational for livestock owners to maintain herd size above the numbers recommended on the basis of carrying capacity estimates. When the vegetation runs out, the animals will move on, weaken, or die, before the range becomes degraded. Moreover, carrying capacity-informed livestock raising is aimed at maximizing beef production, but livestock in Africa and elsewhere is often not raised for its beef only, and, moreover, animals in poor condition can be slaughtered still and some meat and hides recuperated.[13] Finally, livestock not only browses woody vegetation but also facilitates woody vegetation growth, e.g., by grazing (reducing competition of grasses to tree seedlings) and by facilitating seed germination: seeds consumed by cattle that pass through the digestive system are more likely to germinate as long as they are undamaged.[14] The New Rangeland Ecology, therefore, has injected a powerful inclinist correction into what was overwhelmingly a declinist narrative.

[13] Behnke, Scoones, and Kerven (eds.), *Range Ecology at Disequilibrium*; Scoones (ed.), *Living with Uncertainty*. For a New Rangeland Ecology argument criticizing overgrazing crises narratives in the Himalaya, see V. K. Saberwal, "Environmental Alarm and Institutionalized Conservation in Himachal Pradesh, 1865–1994," in A. Agrawal and K. Sivaramakrishnan (eds.), *Agrarian Environments*, pp. 68–85. Researchers have noted that livestock moves on before the vegetation has been depleted, see C. M. McKell, "Multiple Use of Fodder Trees and Shrubs – A World Wide Perspective," in Le Houérou, *Browse in Africa*, pp. 141–149. For an application to North Africa's Mediterranean, see Davis, *Resurrecting the Granary of Rome*, especially pp. 15, 89–123, 170–174.

[14] D. M. Swift, M. B. Conghenour, and M. Atsedu, "Arid and Semi-Arid Ecosystems," in T. R. McClanahan and T. P. Young (eds.), *East African Ecosystems and Their Conservation* (New York: Oxford University Press, 1996), pp. 261–269, and W. D. Pitman, "Contrasts in Current Developments with Tropical Forest Research in Asia and the Americas," in A. Sotomayor-Ríos and W. D. Pitman (eds.), *Tropical Forest Plants: Development and Use* (Boca Raton, FL: CRC Press, 2001), pp. 357–370; J. Piot, "Management and Utilization Methods for Ligneous Forages," pp. 339–349.

In this chapter, as in the previous chapters, the processes of change, rather than an evaluation of the outcome of the process of environmental change in terms of a linear and static declinist or inclinist narrative is foregrounded. In the case of north-central Namibia, however, the declinist narrative continues to be dominant, even as the narrative is opening at its seams.

For example, casual observations about the lack of tree regeneration, such as those made by Ovamboland's foresters in the 1970s, are problematic because tree regeneration is often difficult to detect. Many indigenous trees in southern Africa have a suffrutex habit: the seedling dies back above-ground, but its root system continues to develop. For example, at any given time, a percentage of a population of Transvaal teak (*Pterocarpus angolensis*), a favorite source of livestock browse in north-central Namibia, may be in the suffrutex stage, which can last for seven to ten years. A sample of fourteen Transvaal teak specimens from Ovamboland examined in 1969 demonstrated that the suffrutex stage had lasted nine years on average, having varied from as little as three to as many as twenty-three years.[15]

In fact, the regional foresters' 1970s fears that livestock browsing caused deforestation were not shared universally at the time in north-central Namibia. The region's director of agriculture, e.g., disputed concerns expressed by the Secretary for Bantu Administration in Pretoria in 1972 that overgrazing threatened the region's abundant palm trees. Although the director could not offer any statistics about a decline or increase in the number of palm trees, he emphasized that it was his impression that palm trees suffered little or no damage from livestock browsing and that palm tree regeneration was readily occurring.[16]

Scientists working on tropical Africa have only relatively recently begun to focus in detail on the role of woody vegetation as cattle browse.[17]

[15] Piearce, "Natural Regeneration of Indigenous Trees," p. 115. For Ovamboland, see NAN, OVA 57, Lueckhoff, Report on a visit to South West Africa, November 3–15, 1969, Appendix Regional Forester to Director-in-Chief Department of Bantu Administration Pretoria, Grootfontein, April 3, 1970.

[16] NAN, OVA 49, Director Agriculture Owambo to Sec. Bantu Administration, Ondangwa, February 25, 1972; State Forester to Director Agriculture, Ondangwa, February 21, 1972; Sec. Bantu Administration to Director Agriculture, Pretoria, February 9, 1972.

[17] See Le Houérou, *Browse in Africa*, p. 3, B. H. Walker, "A Review of Browse and Its Role in Livestock Production in Southern Africa," C. M. McKell, "Multiple Use of Fodder Trees and Shrubs – A World Wide Perspective," G. E. Wickens, "The Uses of the Baobab (*Andansonia digitata* L.) in Africa," and H. N. Le Houérou, "The Role of Browse in the Management of Natural Grazing Lands," in Le Houérou, *Browse in Africa*,

In north-central Namibia's semiarid environment, livestock depended heavily on trees and bush as a source of fodder. During the long dry season, nutrient deficiencies were a significant problem and woody vegetation foraging held the key to livestock survival, especially during times of drought. After the end of the rainy season, the grasses in most of north-central Namibia rapidly dried, becoming more difficult to digest, and losing much of their nutritional value. In southern Africa parlance, pastures subject to such conditions are called "sour" veld (savanna).[18] In referring to the end of the rainy season and drought periods, colonial reports on Ovamboland often stated that grazing had become "bad" or "poor" (in terms of nutrition), with a subsequent decline in livestock health. Deficiencies in phosphate, natrium, copper, and protein in dry season forages were pronounced and phosphate deficiency, e.g., was directly linked to a variety of cattle diseases, including the highly prevalent botulism. Chemical analysis of thirty-three grazing samples from Ovamboland collected at the end of the rainy season in April 1969 revealed significant phosphate shortages in all the samples. The authors of the report considered a reasonable grazing phosphate content to be 50–60 ppm (parts per million) but the samples that were drawn from the sandy soils of Ovamboland's "indigenous forest" contained only 20 ppm. Liver analysis of a small number of cattle from Ovamboland also suggested a significant copper deficiency.[19] Commenting on heavy livestock losses in 1941 following a series of drought years, the author of the annual report

respectively p. 12, pp. 141–149, 151–154, and 329–338. See also M. B. Adjei and J. P. Muir, "Current Developments from Tropical Forage Research in Africa," in Sotomayor-Ríos and Pitman, *Tropical Forest Plants*, pp. 331–355, especially 351; P. Frost, "The Ecology of Miombo Woodlands," in B. Campbell (ed.), *The Miombo in Transition*, pp. 36–38; Behnke, Scoones, and Kerven (eds.), *Range Ecology at Disequilibrium*; and W. Bayer and A. Waters-Bayer, "Forage Alternatives from Range and Field: Pastoral Forage Management and Improvement in the African Drylands," in I. Scoones (ed.), *Living with Uncertainty*, pp. 58–78.

18 NAN, NAO 18, Monthly Reports Ovamboland, February–March 1928; NAO 101, Agricultural Survey of Ovamboland, Windhoek, October 26, 1947; NAO 62, Agricultural Report Ovamboland, Omafu, November 30, 1953.

19 NAN, AHE (BAC) 1/352, Annual Report Agriculture Ovamboland, 1968; OVA 40 f. 6/5/2, Deputy Head Veterinary Services to Director-in-Chief Ovamboland, Windhoek, August 29, 1969, and Appendix Chemical composition of grazing samples in Ovamboland, April 23, 1969, and Ovambo Government, Report Verhoging van die Beesvleis produksie in Ovambo, August 5, 1971; OVA 56, Sec. Bantu Administration to Director Agriculture Ondangwa, Pretoria, June 11, 1971; OVA 46, Agriculture to Sec. Bantu Administration Pretoria, Ondangwa, November 17, 1971; OVA 57, Forest Inventory Indigenous Forests Ovamboland, Le Roux to Sec. Agriculture, Ondangwa, November 5, 1976; OVA 6, Annual Report Veterinary Services Ovamboland, 1978/1979.

for Ovamboland commented that goats could get by but "[a]s for large stock it is hard to understand on what the Ovambo beast subsists at the present time."[20]

The answer was trees and bush. Typically, late in the rainy season and during the dry season, reports mentioned "the bush," "bush," or "bush country," as the major sources of livestock fodder, although the word "grazing" continued to be used where "browsing" would have been more accurate.[21] For example, a July 1939 report commenting on the situation early in the dry season stated:

> [i]n the Eastern Ondonga-Ukuanyama bush the grazing on the whole is good, and once away from the Omusati [*omufyaati* or mopane] belt there is an abundance of Efufe [*ofufe* or sand camwood] and Ompanda [*omupanda* or apple-leaf] bush, the foliage of which is greatly sought after by both cattle and horses and which is a real stand-by during years of drought.[22]

In the 1993 OMITI survey, respondents identified mopane, sand camwood (*Baphia massaiensis*), apple-leaf (*Lonchocarpus* spp.), silver cluster leaf, palm, marula, birdplum, and fig as major sources of browse. Minor sources of browse included buffalo thorn, baobab, Transvaal teak, jackalberry, and camel thorn (*Acacia erioloba*). Some, including palm, were relatively more important browse for goats than for cattle and the seed pots or fruits of camel thorn, fig, and palm were also significant sources of goat fodder.[23] Other tree species identified as sources of livestock forage included sickle bush (*Dichrostachys cinerea*) and tamboti, both of which have been identified as bush encroachers.[24]

Early in 1960, most of Ovamboland suffered from a severe drought. The only green vegetation consisted of trees and bush "on which the

[20] NAN, A450 vol. 7, Annual Report Ovamboland 1941.

[21] NAN, NAO 17, NCO to Clarke, Ondangwa, July 17, 1933; NAO 23, NCO to O/C Oshikango, Ondangwa, July 15, 1939, and O/C Oshikango to Wessels, [Oshikango] July 17, 1939; NAO 18–20, Monthly Reports Ovamboland, November–December 1930, September–October 1936, September 1938; NAO 60–61, Quarterly Reports Ovamboland April–September 1948, April–December 1949, October–December 1953; NAO 41, Famine Relief Report Ovamboland, October 1930.

[22] NAN, NAO 9, O/C NAO to NCO, Ondangwa, July 26, 1939.

[23] OMITI survey 2.1.26 (cattle) and 2.2.7 (goats).

[24] Rundu Forestry, file Protected Species, List of trees with botanical names and local names (Kavango and Owambo names); Rodin, *The Ethnobotany of the Kwanyama Ovambos*. See also Kreike, "Architects of Nature," Appendix 2: Fodder Trees and Bushes by Latin-Ovambo names, pp. 273–274.

livestock largely lived."[25] Mopane bush and other woody species that cattle and goats feed on become green before most grasses.[26] The same is true in the West African Sahel, where the semiarid environment is comparable. Many Sahelian trees and shrubs develop new leaves several weeks before the onset of the rainy season. The leaves are the only green vegetation available to the livestock and wildlife, which otherwise survive on dry grass that has little nutritional value.[27]

Research from the West African Sahel has demonstrated that in dry areas, woody vegetation browse contributed at least 5 percent of livestock fodder intake during the rainy season and early in the dry season. During the peak of the dry season, the woody vegetation intake rose to 15–25 percent. Dry grass continued to provide most of the animals' energy. Although such browsers as goat, eland, impala, kudu, and elephant can sustain themselves on a diet that consists exclusively of woody vegetation, cattle cannot; an exclusive browse diet does not meet cattle's energy requirements.[28] Despite the considerable seasonal variation in the nutritional value of tree and bush browse, it is much more nutritionally stable than grasses. Research in South Africa indicated that although the crude protein content of the widely occurring mopane plant fell between January and May (during the rainy season), calcium and phosphorous levels in the sampled mopane leaves actually increased.[29]

Browse in the Angolan-Namibian border region was (and is) as critical to cattle during the dry season as it is in the West African Sahel. For example, the fresh shoots and young leaves of mopane, sand camwood, apple-leaf, palm, fig, silver cluster leaf, marula, birdplum, and camel thorn are an important source of browse later in the dry season.[30]

But only after unsuccessful trials with exotic fodder bush in the early 1960s did a forester in the late 1960s recommend research trials with

[25] NAN, BAC 123, Chief Bantu Commissioner SWA to Sec. Bantu Administration, Windhoek, February 8, 1960.

[26] Personal observations, 1991–1993.

[27] Le Houérou, *The Grazing Land Ecosystems*, pp. 49–50, 82.

[28] Le Houérou, "The Role of Browse in the Management of Natural Grazing Lands," in Le Houérou, *Browse in Africa*, pp. 329–338.

[29] Walker, "A Review of Browse and Its Role in Livestock Production in Southern Africa," pp. 16, table 9. Research in South Africa suggested that deciduous trees on poor soils remove half of the phosphorous and potassium from their leaves before they shed them but none of the calcium or magnesium. Deciduous trees on richer savanna soils do not translocate these elements before they shed. See Scholes and Walker, *An African savanna*, p. 108.

[30] OMITI survey, 2.1.26.

local indigenous fodder trees, including mopane.[31] Subsequent trials with indigenous cattle and forages in the late 1960s and 1970s revealed that the bush and the tree leaves in north-central Namibia remained nutritious much longer than grasses and herbs and that they contained much higher concentrations of such essential minerals as, e.g., phosphate and copper.[32]

A 1979 report on indigenous cattle in southern Africa emphasized that

> [s]anga cattle are also browsers... they make use of foliage in their diets and therefore withstand droughts in bushveld better than any grazers since they often obtain sustenance from green leaves of a number of tree species or fallen leaves picked up beneath the tree.[33]

The hardy mopane emerged as, by far, the most important source of browse in north-central Namibia. A late 1960s report on local trials at Ogongo Agricultural College involving a plot of "mopane pasture" showed mopane as the dominant tree and bush on the plot. The weight of dry vegetative matter on the mopane pasture plot was found to be 1,160 kilo per hectare while a trial plot of bushwillow (*combretum*) pasture only contained half that amount. Although researchers found the tree and bush cover of the bushwillow "pasture" to be high and its relative grass cover to be low, they nevertheless concluded that bushwillow pasture could support one cattle unit (CU) per ten hectares. Because the dry vegetative matter mass measured on mopane "pasture" was double that which was measured on bushwillow pasture, mopane pasture should be able to sustain a considerably higher density of cattle than bushwillow pasture.[34]

31 NAN, BAC 131, Agricultural Officer Ovamboland to Bantu Commissioners Ondangwa and Oshikango, January 28, 1957; BAC 132, Agriculture Officer to Bantu Commissioner, Grootfontein, July 30, 1961: Report on Work on Okatana Irrigation Scheme, July 30–September 11, 1961; Chief Bantu Commissioner to Bantu Commissioner Grootfontein, Windhoek, August 3, 1961; Agriculture Officer Grootfontein to Bantu Commissioner, Grootfontein, July 31, 1962, and Chief Bantu Commissioner to Bantu Commissioner Ondangwa, September 4, 1962; OVA 57, Dr. H. A. Lueckhoff, Report on a visit to South West Africa, November 3–15, 1969, Appendix Regional Forester to Director-in-Chief Bantu Administration Pretoria, Grootfontein, April 3, 1970.

32 Kreike, "Architects of Nature," pp. 132–133.

33 NAN, OVA 26, H. Hamburger, C. W. B. Amstrong, and J. Swanepoel, Adaptability and Reproductive Efficiency: The Value of Indigenous Sanga Cattle in the National States of South Africa and Namibia, RSA Department of Co-operation and Development [1979?].

34 NAN, OVA 44, Opperman and Prinsloo, Botanical Survey and Physical Planning Orongo Trial Area, Appendix to Opperman to Director Agriculture, Windhoek, July 28, 1969. See E. Kreike, "De-Globalization and Deforestation in Colonial Africa: Closed

The 1975/1976 progress report about the Ovamboland cattle development project noted that the green mopane bushland that supplied forage late in the dry season and early in the rainy season was superior to forage contained in the dry *oshana* pastures (the grasslands covering the seasonal watercourses).[35] A 1979 report reiterated the importance of mopane as a dry season source of fodder.[36] Significantly, the "mopane pastures" described were actually primarily covered with mopane bush, rather than grass. The less-nutritious *oshana* pastures located in the low-lying areas of the floodplain consisted of grasses and herbs with little or no tree or bush vegetation. Further 1970s research on *oshana* pastures concluded that they were abundant in central Ovamboland, that they received large amounts of floodwater in addition to local rainfall, and that the pasture was "sweet," a South African term that highlights that the pasture remained nutritious after the vegetation had dried out.[37] This assessment, however, is probably in error; *oshana* pasture is more likely "sour" grazing. After the *oshana* flood channels dry out, its tall grasses quickly turn yellow (although later than the grasses on the ridges) and they contain very little nutritious woody vegetation.

In terms of its dominant presence, usefulness, tenacity, and symbolism, and its invasive properties, mopane is in fact very much the oak of Africa (or the oak is the mopane of the West).[38] Like oak in temperate and subtropical Europe and North America, mopane is an important source of browse throughout much of southern Africa.[39] Colonial experts identified mopane repeatedly as an important livestock fodder plant and trials at the north-central Namibia Ogongo Agricultural College highlighted the fodder potential of mopane forage.[40] Again, like oak, mopane leaves are

Markets, the Cattle Complex, and Environmental Change in North-Central Namibia, 1890–1990," *Journal of Southern African Studies*, 35 (2009), pp. 81–98.

35 NAN, OVA 56, F.J. van der Merwe (Veterinary Studies, University of Stellenbosch), Progress Report 1975/1976: Research into the Observed Low Meat Production Potential of the Indigenous Cattletype in Owamboland.

36 NAN, OVA 26, Owambo Livestock Practices, appendix to Sec. Agriculture to Sec. Bantu Administration, Pretoria, Ondangwa, March 10, 1977.

37 NAN, OVA 43, Joubert, Final Report: Research on the Productive Potential and Best Use of Oshana Pasture Veld in Owambo, Appendix to Sec. Plural Relations to Sec. Agriculture Ovambo Government, Pretoria, June 5, 1979.

38 For Zimbabwe, see Mushove and Makoni, "Coppicing Ability of *Colophospermum mopane*," pp. 226–230.

39 Walker, "A Review of Browse," pp. 12–14, 16–17, 22.

40 NAN, OVA 56, van der Merwe, Progress Report 1975/1976: Research into the Observed Low Meat Production Potential of the Indigenous Cattletype in Owamboland. OVA 44, Opperman and Prinsloo, Botanical Survey and Physical Planning Orongo Trial Area,

high in tannins. In fact, young mopane leaves may be so high in tannins that they are avoided by most ungulates; when they are dry, however, cattle and wild ungulates eat them readily.[41] With mopane as the main browsing species in Ovamboland and across semiarid southern Africa, its prevalence in Ovamboland and throughout the region not only serves to highlight the tensions inherent in projecting the relationship between cattle and environmental degradation in linear and mechanical terms but also raises the possibility that the abundance of mopane may, in part, be a product of its importance and intensive use as browse. Dry mopane leaves are most palatable to livestock during the height of the dry season; the high tannin levels in the green leaves may protect mopane during its growing season. Under such conditions, mopane may thus have a competitive edge over more palatable species whose leaves are consumed green, including marula and birdplum. Marula, birdplum, and other fruit trees are a favorite livestock browse and it is perhaps no surprise that marula and birdplum in north-central Namibia occur only when and where they are protected by humans, on fenced-in fields and farms in the villages.[42]

Moreover, it is significant that mopane and other browsed trees and bushes that occur in north-central Namibia are elsewhere considered to be dangerous invasive species that cause bush encroachment. Despite the abundance of mopane and other invasive species, colonial and postcolonial experts did not identify bush encroachment as a significant environmental threat in north-central Namibia. This gives pause because bush encroachment is a major problem in the so-called commercial farming areas across southern Africa, as well as in semiarid Africa in general, in the Mediterranean, and in the chaparral regions of the Southwestern United States.[43] Like oak in the Southwestern United States and in the Mediterranean basin, mopane was and is a nearly indestructible invasive species. In 1930s and 1940s Ovamboland, mopane was singled out

Appendix to Opperman to Director Agriculture, Windhoek, July 28, 1969; OVA 26, Owambo Livestock Practices, Appendix to Secretary Agriculture to Secretary Bantu Administration Pretoria, Ondangwa, March 10, 1977.

41 Walker, "A Review of Browse," p. 10. But the responses to the OMITI survey in the previous discussion include young mopane leaves and even its shoots as a source of browse very late in the dry season.

42 Erastus Shilongo, interview by the author, Olupanda (Okalongo, Namibia), June 21, 1993. In general, see Chapter 2.

43 For bush encroachment in the Southwestern United States, see Williams et al., "Agroforestry in North America and Its Role in Farming Systems," pp. 9–84 (especially 30 on mesquite).

as an encroaching bush in colonial clearings, including those made to demarcate the Angolan-Namibian boundary as well as dirty roads and airstrips. Repeating a refrain in many reports, the quarterly report for the Oukwanyama district covering the first months of the rainy season of 1949 complained that "[t]he majority of the roads are overgrown with grass and Mopani [mopane] and need clearing."[44]

By 1959, bush encroachment had become such a severe environmental problem to white settler farmers in colonial Namibia that the Congress of the South West African Agricultural Union appealed to the administration to declare it a national threat and to take the lead to combat it.[45] A 1967 report on bush encroachment noted that the administration had heeded the call, but without any success. Chemical eradication with aboricides had been effective in suppressing invasive bush but was uneconomical. Mechanical clearing with bulldozer blades made matters worse, not only because it left soils exposed to erosion but also because wherever the feared black thorn was successfully removed it was replaced by the even more tenacious sickle bush.[46] In 1972, 8,000,000 ha in the northern and central white settler-farming districts of colonial Namibia were listed as bush encroached. More than half of the total affected surface area was considered to be "highly" bush-encroached. In the northern districts (just south of Ovamboland), the lands with the best soils were also the worst affected by bush encroachment.[47] By the 1980s, 40,000,000 ha of land in Namibia and South Africa combined were considered to be bush encroached.[48] Today, bush encroachment continues to be an enormous environmental and economic problem across Namibia and South Africa, especially in the former white settler "commercial" farming regions.

Reports from the 1960s and 1970s include mopane as a major bush encroaching species in the white farming areas directly south and

[44] NAN, NAO 60, Quarterly Report Oukwanyama, January–March 1949. The history of bamboo, although categorized as a grass rather than a woody species, is similar in many respects to that of mopane. It has been widely used and managed, yet is mostly seen as a wild invader. For bamboo as a wild weed, see Gadgil and Guha, *This Fissured Land*, pp. 171–172.

[45] NAN, AGR 562, Congress South West African Agricultural Union, resolution no. 31, December 2, 1959.

[46] NAN, AGR 563, D. J. Louw, "Bosindringing in Suidwes-Afrika," May 19, 1967, and Appendix to Pasture research Officer to DOW Chemical Africa, Windhoek, August 19, 1964.

[47] NAN, AGR 563, acting Director [Nature Conservation?] to J. van Wyk, Windhoek, August 18, 1972.

[48] See I. A. W. MacDonald, "Man's Role in Changing the Face of Southern Africa," in B. J. Huntley (ed.), *Biotic Diversity in Southern Africa: Concepts and Conservation* (Cape Town: Oxford University Press, 1989), pp. 51–77 (especially 68).

southeast of Etosha Park and Ovamboland. Other dangerous bush encroachers included not only black thorn and sickle bush but also red umbrella thorn (*Acacia refisciens*), blue thorn (*Acacia erubescens*), plate thorn (*Acacia fleckii*), silver cluster leaf, and tamboti.[49] Climate and vegetation in the white settler-farming districts south of Etosha and Ovamboland were very similar to that in Ovamboland. Many of the woody vegetation species in the Tsumeb-Grootfontein-Otavi region and the Outjo region occurred in Ovamboland, including not only marula, Lowveld cluster-leaf (*Terminalia prunioides*), shepherd's tree (*Boscia albitrunca*), red bushwillow, and wild raisin bush, but also the bush encroachers listed previously.[50]

Colonial research concluded that once established, encroached bush was virtually impossible to eradicate. More unsettling, however, was the finding that white settler farming (i.e., modern commercial cattle ranching) was largely responsible for the problem. In a 1961 questionnaire distributed for a large pasture research project the Director for Agriculture of colonial Namibia responded that "[bush] [e]ncroachment is associated with European settlement and cattle farming." The extensive felling of trees in the white settler farming region of northern Namibia was listed as one cause of bush encroachment: "where the trees were cut some years ago for pit props in the mines in the Tsumeb area, the cut[t]ing resulted in a very dense coppice growth."[51]

Even more damning and potentially subversive to the scientific underpinning of the colonial project as a whole, however, was that the *prohibition* in the white commercial farming areas of two of the most maligned forms of African land use (burning and goat keeping) proved to be the key causes of bush encroachment. Acknowledging the finding that white settler farmers were responsible for severe environmental degradation could have deracialized and subverted colonial science. But the colonial scientific establishment persisted in condemning seasonal burning and goatkeeping as primitive and irrational environmental evils. Even as their research reports grudgingly acknowledged the value of goat browsing

49 NAN, AGR 562, report "visit to Mr. De Wet Outjo regarding machine for bush eradication," April 4, 1963, and Director of Agriculture to Chief Pasture Research Officer Salisbury, Windhoek, March 29, 1961, and Appendix; AGR 563, acting Director [Nature Conservation?] to J. van Wyk, Windhoek, August 18, 1972.

50 NAN, AGR 562, Director of Agriculture to Chief Pasture Research Officer Salisbury, Windhoek, March 29, 1961, and Appendix.

51 NAN, AGR 562, f. 68/19 (I), Director of Agriculture to Chief Pasture Research Officer Salisbury, Windhoek, March 29, 1961, and Appendix. On fire exclusion and bush encroachment on white farms, see also Scholes and Walker, *An African Savanna*, pp. 46–48 and 111–125.

and seasonal burning in preventing bush encroachment, colonial scientists neither advanced goats or fire as a solution nor disseminated their findings beyond a small circle of researchers.

Cultivating the Bush with Goats

In his response to the 1961 questionnaire on pasture research, the Director for Agriculture hesitantly had suggested that goat browsing might contribute to checking bush encroachment. He explained that trials were under way at colonial Namibia's Omatjenne Experimental Farm to assess whether goats browsed bush encroaching species and with what outcomes. He noted that in contrast to the situation in white settler farming regions, bush encroachment was not a problem in the Damaraland African reserve even though it was located adjacent to bush encroached white farmlands. The Director of Agriculture hypothesized that goats debarked the most important bush encroaching species in the reserve during the dry season, including the pernicious black thorn (*Acacia mellifera*) and that this acted as a check on the species' spread.[52]

The Director of Agriculture omitted mention of the fact that his staff at Omatjenne Experimental Farm had come to the startling conclusion that goats prevented bush encroachment. The trials demonstrated that early during the Summer, that is during the dry season, when grass was scarce, goats ate the bark of the dangerous bush encroachers black thorn and white thorn (*Acacia volkii*), killing the plants.[53] During the 1950s and 1960s, the expulsion of African squatters' livestock (especially goats) and the fencing of white-owned commercial farms effectively excluded domestic and wild browsers – notably goats – from the white-owned farmlands. As exotic cattle increased pressure on grasses and herbs, the bush was left unchecked, creating favorable conditions for bush encroachment. A 1964 report based on a large research trial that included aerial spraying (probably of arboricides) in the Grootfontein-Tsumeb white settler farming region concluded that competition by grasses was the most effective way to limit bush encroachment. Pasture management thus emerged as the key to combating bush encroachment.[54]

[52] NAN, AGR 562, Director of Agriculture to Chief Pasture Research Officer Salisbury, Windhoek, March 29, 1961, and Appendix.

[53] NAN, AGR 562, J. Joubert, "Progress Research on Fighting Bush Encroachment," March 9, 1960.

[54] AGR 563, D. J. Louw, "Bosindringing in Suidwes-Afrika," May 19, 1967; AGR 563, Appendix to Pasture Research Officer to DOW Chemical Africa, Windhoek, August 19, 1964.

Despite the evidence from indigenous practices in Damaraland and the scientific trials at Omatjenne Research Station, a 1972 report downplayed the potential role of goats in combating bush encroachment. It argued that no amount of goats could reclaim an area once it was bush encroached because most of the vegetation would be physically out of their reach. The report's author concluded that at best, goats could slow down the regeneration of bush vegetation because the animals kept the young bush short or killed it.[55] When the Omatjenne trials ended in 1973, however, the results clearly proved that goats not only prevented bush encroachment but that their browsing actually improved grass cover. Goats, however, were still considered to cause overgrazing because they also fed on young grass when bush forage was in short supply. The recommendation to white settler farmers of the multiyear Omatjenne trials on goats and bush encroachment was not to shift to raising goats but to take up game ranching with browsing species.[56] A study published in Cape Town in the late 1980s – toward the end of the colonial Apartheid era – recommended cattle as the preferred grazers in savanna vegetation but added "[that] goats should usually be introduced here to provide the browsing component... to optimize red meat production. Browsers may also be introduced to manipulate the grass [to] bush ratio."[57]

The reluctance to acknowledge that goats could, under specific circumstances, make a positive contribution to land management is not surprising. In general, goats had acquired a bad reputation as harbingers of deforestation and desertification. The association of deforestation and goats dates back to at least the early days of the development of the environmental crisis concept in the eighteenth century.[58] Mediterranean goats are considered an environmental scourge. Colonial officers in Kenya identified goats as the main cause of a perceived degradation of the Great Rift

55 AGR 563, acting Director [Nature Conservation?] to J. van Wyk, Windhoek, August 18, 1972.

56 NAN, AGR 541, P. A. J. Brand, "Use of Game for Extensive Livestock Ranching," n.p., n.d. [in top margin: "article from Swalarama 1973"].

57 N. M. Taiton, P. J. K. Zacharias, and M. B. Hardy, "The Contribution of Veld Diversity to the Agricultural Economy," in B. J. Huntley (ed.), *Biotic Diversity in Southern Africa*, pp. 107–120 (especially 112). One of the sources cited was a 1977/1978 annual report of colonial Zimbabwe's Division of Livestock and Pasture Management; another source was B. H. Walker, "A Review of Browse and Its Role in Livestock Production in Southern Africa," pp. 7–24.

58 Grove, *Green Imperialism*, pp. 95–120. See also M. Williams, *Deforesting the Earth: From Prehistory to Global Crisis* (Chicago: University of Chicago Press, 2003), p. 83, and Radkau, *Nature and Power*, pp. 23–25.

Valley pastures.[59] Colonial officials in north-central Namibia had similar views. Annual reports in the early 1940s singled out Ovamboland's goats as "the most miserable species in South West Africa if not in the Southern Hemisphere" and described the animals as an environmental and health "curse."[60]

Information on goats in north-central Namibia is rather scarce, especially in the colonial reports where they are nearly invisible. Goats were common in the northern Ovambo floodplain in modern Angola in the late 1880s and early 1900s. Households kept their goats in separate smallstock kraals and young boys herded the goats during the rainy season to prevent them from damaging crops. As a rule, smallstock remained at the villages throughout the year, but during the extreme drought conditions of September 1927 even small stock were herded across long distances to cattleposts. During the 1929–1931 Famine of the Dams, goat milk was an important alternative to cow milk because the cattle remained at cattleposts far away from the villages. Because they were browsers and because bushes and trees sprouted new leaves before grasses and herbs, goat milk was a critical source of food early during the rainy season when little food other than stored grain and dried foods were available.[61]

59 In his influential work, Thirgood identifies goats and sheep grazing as the major source for environmental degradation in the Mediterranean as a whole and Cyprus in particular. See, Thirgood, *Cyprus*. Hughes also highlights the negative impact of smallstock on the Mediterranean environment; see J. Donald Hughes, *The Mediterranean: An Environmental History* (Santa Barbara, CA: ABC Clio, 2005), p. 93, and Hughes, *Pan's Travail*, pp. 78–79. Both Hughes and McNeill argue that the goat in the Mediterranean is not the major *cause* of deforestation but argue that it prevents forest recovery, see Hughes, *Pan's Travail*, pp. 78–79, and McNeill, *The Mountains of the Mediterranean World*, pp. 86, 92–93, 128–129, 244–245, 276, 277, 311–325, 336–338. The bush encroachment research at Omatjenne does not contradict the latter: bush encroachment after all is a form of reforestation, albeit with species considered to be undesirable. Davis, on the basis of her research in the North African western Mediterranean, argues that the declinist narrative was a French colonial invention. See Davis, *Resurrecting the Granary of Rome*. On Kenya, see Anderson, *Eroding the Commons*, pp. 126–156. Tiffen et al. ventured that the "maligned goat" might have contributed to the control of bush encroachment in the Machakos district of Kenya; see Tiffen et al., *More People, Less Erosion*, p. 127.

60 NAN A450 vol. 7, Annual Reports Ovamboland 1940 and 1941.

61 NAN, NAO 104, Anderson to Hahn, Jordan diary, pp. 519–539. AVEM, RMG 2599 C/i 19, Bernsmann, Omburo, January 6, 1892 (livestock kraals). On goats, see Wulfhorst, *Shiwesa, ein Simeon aus den Ovambochristen*, p. 2 and CNDIH, Avulsos, Caixa 3439 "Cuamato," Ribeiro da Fonseca (Tenente da Cavalaria), "Relatório do reconhecimento feito pelo 20. esquadro de dragões á região da Inga, Oucuancua e Dombondola," Cuamato, September 26, 1913. For oral histories, see, e.g., Paulus Nadenga, interview

Colonial statistics on the number of goats are largely based on guesstimates. Even as the number of goats increased in the long run from 80,000 in 1925 to 450,000 in 1960, their numbers actually declined at times. For example, the number of goats fell from 168,000 in 1935 to 103,000 in 1942; the goat population more than halved between 1960 and 1966; and during the late 1960s, 1970s, and 1980s, the numbers fluctuated up and down between 300,000 and 400,000. The cattle population was higher, ranging between 400,000 and 500,000 during the 1960s–1980s. Although cattle statistics for those years are more reliable because they can be checked against the number of vaccination doses distributed by veterinary officials, equivalent records for dipping goats are rare. Counts for dipped goats are known only for 1968 and 1969. In both cases, only part of the goat population was dipped. The 1969 figure of 229,000 dipped goats included only the large Oukwanyama district and three of the smaller western districts and excluded the Ondonga and Uukwambi districts. But the dipped number suggests that the estimate of 300,000–350,000 animals is reasonable.[62] The year-to-year fluctuations are not entirely statistical artifacts; qualitative information confirms that goat populations were decimated by drought and disease and the estimates

by author, Oshomukwiyu (Namibia), April 28, 1993. Famine of the Dams: NAO 18, Monthly Reports Ovamboland, September 1927 and November–December 1930.

62 Compiled from NAN, A450 vol. 7, Annual Reports Stock 1935 and Annual Reports Ovamboland 1935, 1942; BAC 133, Agricultural Reports Ovamboland 1955/1956, 1956/1957; BAC 131, Agricultural Officer Ovamboland to Bantu Commissioners Ondangwa and Oshikango, January 28, 1957; AGR 44, f. 6/2/1, Zschokke, Ondangwa, September 2, 1958; WAT 1, "Cattle in Northern Native Territories," June 30, 1960, Ovamboland; AHE (BAC) 1/352, Annual Reports Agriculture Ovamboland for 1966 and 1968; OVA 9, Statistics 1967, Appendix to Director-in-Chief Economic Affairs to Director Agriculture, Ondangwa, March 25, 1969, and Chief Agricultural Official to Director Agriculture and Works, Ondangwa, June 25, 1969, "Regarding Questionnaire" and Director Agriculture and Forestry to Director-General Department of Cooperation and Development Pretoria, [Ondangwa], May 5, 1981; OVE 10, table 6.1.3; OVA 40, Ovambo Government, Report "Verhoging van die Beesvleis produksie . . . ," August 5, 1971, and questionnaire appended to Secretary Economic Affairs to Secretary of Agriculture and Forestry, Ondangwa, November 10, 1973; OVA 26, "Owambo Livestock Practices," Appendix to Secretary Department of Agriculture to Secretary Department of Bantu Administration, Ondangwa, March 10, 1977; AGR 897, De Basson, "Numbers Ovamboland," February 24, 1963; OVA 6, Annual Report Veterinary Service Owambo 1975–1976; OVA 26, "Owambo Livestock Practices," Appendix to Secretary Department of Agriculture and Forestry to Secretary Department of Bantu Administration and Development, Pretoria, Ondangwa, March 10, 1977; Kreike "The Ovambo Agro-Silvipastoral System," p. 39, table 1. The 1966–1975 figures from the 1977 report seem sanitized and may have been derived from a formula that reflected presumed drought and disease losses of cattle.

actually seem to have been adjusted to account for such losses. For example, the previously mentioned drops in the goat numbers during the early 1940s and 1960s were the result of severe droughts that caused high goat mortality. Officials estimated losses at 20 percent of the total herd in the drought year of 1941 alone, and in 1979/1980, 20,000 goats reportedly succumbed to a combination of disease and drought.[63] In addition, in contrast to cattle that were chiefly slaughtered for special occasions or during droughts, the consumption of goat meat and the sale of goats was much more common.[64] With high annual fluctuations in goat numbers, their impact on the landscape was also highly variable in time.

Moreover, goats were not evenly distributed across the region in spatial terms and the ratio between human and goat populations fluctuated dramatically. The most populous districts of Oukwanyama and Ondonga had also the highest numbers of goats. Although Oukwanyama in 1963 was by far the goat-richest district with 160,000 animals, Ondonga had taken over that role by 1976. Between 1963 and 1971, a period of only eight years, the number of goats in Oukwanyama declined by half. In the smaller western districts of Ongandjera, Ombalantu, Uukwaluthi, and Onkolonkathi the goat populations oscillated between 10,000 and 30,000 head.[65]

[63] NAN, BAC, Agricultural Reports Ovamboland, 1955/1956 and 1956/1957; NAO 103, Census of Agriculture Ovamboland, 1945/1946, and NCO to Secretary SWA, Windhoek, March 8, 1946; BOS, Agricultural Officer Ovamboland to Native Commissioners Ondangwa and Oshikango, [Ondangwa], August 17, 1956; OVA 6, Annual Report Agriculture Owambo, 1979/1980; A450 vol. 7, Annual Report Ovamboland 1941.

[64] NAN, OVA 9, Director Agriculture to Department of Cooperation Pretoria, [Ondangwa], May 5, 1981; OVA 26, "Owambo Livestock Practices," Appendix to Secretary Department of Agriculture to Secretary Department of Bantu Administration, Ondangwa, March 10, 1977. These figures are estimates.

[65] AGR 897, De Basson, "Numbers Ovamboland," February 24, 1963; AHE (BAC) 1/352, Annual Reports Agriculture Ovamboland for 1964, 1966, and 1968; OVE 10, table 6.1.3. For alternative 1967 figures, OVA 49, Agricultural Statistics 1967, appendix to Chief Director Department of Economic Affairs to Director Agriculture, Ondangwa, March 25, 1967. For the 1977 figures, see OVA 26, "Owambo Livestock Practices," Appendix to Secretary Department of Agriculture and Forestry to Secretary Department of Bantu Administration and Development, Pretoria, Ondangwa, March 10, 1977. For differentiation by districts, see OVA 56, State Veterinarian to Director Agriculture, Ondangwa, June 7, 1971, and OVA 6, Annual Report Veterinary Service Owambo 1975/1976. The 1971 figures were at least in part based on goat dipping reports. For 1969 dipping statistics, see NAN, OVA 56, Dip reports Oukwanyama and Ombalantu, October–November 1969, and Uukwaluthi and Ongandjera, July–September 1969, Appendices to State Veterinarian to Director Agriculture, Ovambo People's Government, Ondangwa, January 7, 1970.

The impact of goats on the landscape as a whole was smaller than that of cattle: they consumed less forage per animal, their numbers were lower overall, and their numbers fluctuated more rapidly from year to year. Seasonally, spatially, and environmentally, the impact of goats was much more concentrated than that of cattle. Whereas cattle were mainly grazers, goats were mainly browsers. Although herdsmen drove cattle to remote cattleposts during the dry season, goats, other smallstock, and donkeys and horses were kept near the villages, where they consequently largely impacted on the village environments. During the rainy season, boys (sometimes as young as four to six years of age), girls, or women herded goats and calves to keep them from damaging crops. During the dry season, however, smallstock roamed freely in and near the villages.[66]

In 1993, in half out of a sample of 258 households, children herded the goats during the rainy season; in only 35 households, the goats were not herded at all during the rainy season. During the dry season, the overwhelming number of households did not herd their goats; only one out of every ten households reported that they continued to herd their goats. Unguarded during the dry season, goats roamed over much larger distances of up to five kilometers (three miles) and beyond in search of forage and water. During the rainy season, when they were herded and when forage and water was much more abundant, goats remained much closer to home, often within a kilometer (three quarters of a mile) of their owners' homestead.[67]

[66] Paulus Nadenga, interview by the author, Oshomukwiyu (Namibia), April 28, 1993; NAN, NAO 18–20, Monthly Reports Ovamboland, February and September 1927, November 1928, June–August and November–December 1929, May–June and November–December 1930, September–October 1932, October–December 1937, and July–August 1938; NAO 60, Quarterly Report Ovamboland October–December 1952; NAO 90, Statement Haipinge Makanda, Ondangwa, April 22, 1950; NAO 13, NCO to Secretary SWA, December 12, 1934, and NCO to Secretary SWA, Ondangwa, January 18, 1935, and Alho (FMS) to NCO, Olukonda, January 15, 1935, and Statement Lucas Shilunga, September 17, 1935; NAO 45, Statement Andreas ya Zacharias, Onayena, January 18, 1940; NAO 10, Minutes of a Special Meeting, June 19, 1931; A450, 9, Tobias, October 1938; NAO 45, Statement Matteus Angula, September 14, 1939; NAO 91, Chief Kambonde to NCO, Okaroko, June 14, 1953, and Statement Thomas Kupila, May 5, 1954; NAO 92, Council Headmen Ombalantu to NCO, October 9, 1954; NAO 46, Statement Nangombe Nute, Appendix to NCO to Attorney-General, Ondangwa, June 4, 1941; A450 vol. 12, SWA Commission, Minutes of Evidence vol. 12, Ukauluthi, August 13, 1935, evidence Hahn, p. 654; BAC 45, Minutes Tribal Meetings Oukwanyama, November 8–22, 1957; WAT 146, Water Affairs to Director Windhoek, Ondangwa, October 19, 1961; MacDonald Diary, pp. 2, 9; Kreike, "Recreating Eden," chapter 8.

[67] OMITI survey, 2.2.4–5, 2.2.6.0–1.

Because floodplain villages were very close together, goat browsing heavily shaped the floodplain vegetation. Goats had much less impact on the western and southern margins of the floodplain (where villages were few) or in eastern Ovamboland (beyond the floodplain) where villages were located at greater distances from one another. Suggestive of the heavy impact of browse in and around the villages is the fact that the regeneration of two of the most popular browsing species in north-central Namibia's villages, the marula and birdplum fruit trees, is confined to fenced locations that prevented access by livestock.[68] The other fruit trees that are a source of browse, including palm, fig, and jackalberry, are not typically fenced, but they are protected by the herders who accompany the livestock. The fodder species of sand camwood and apple-leaf also receive protection. Fencing and herding thus affected how browsers shaped the tree and bush vegetation of Ovamboland.

The use of browsers and other management practices also impacted the bush encroachers that pervaded the vegetation of north-central Namibia. Mopane and the other species that have overgrown farm and ranchlands in Namibia's "commercial" farming districts all occur in Ovamboland, but they are either heavily used as browse, as browse and construction material, or – in the case of the thorny acacias – are used extensively as fencing materials and thus do not cause bush encroachment. Some of the most tenacious bush encroachers in the historical white settler farming regions *south of* Ovamboland, including mopane, sickle bush, silver cluster leaf, and tamboti, are important sources of browse *in* Ovamboland. Next to browsing and fencing, the use of fire has played an important role in preventing bush encroachment in north-central Namibia.

Cultivating with Fire

Until it was suppressed in the twentieth century and before the emergence of massive construction projects (dams), industrial pollution and global climate change, and modern weapons of mass destruction, fire was probably the most dramatic creator and destructor of environmental infrastructure globally. Cooking made a much wider range of foods digestible, improving diets and human health and facilitating population growth. Burns shaped flora and fauna, facilitated mining and metallurgy, provided weapons of war (including "firearms"), and steam and internal combustion engines provided (and continue to provide) the energy

68 See Chapter 2.

to sustain human society.[69] More than any other pre–twentieth-century tool, however, fire was a two-edged sword because it was – and is – very difficult to control. The cumulative fires that sustain human society are slowly "cooking" our planet, resulting in climate change. Moreover, the extent to which landscape-defining fires originated as a result of human agency is sometimes unclear, and even when and where humans caused the fire, it may have been unintentional in its origins and unpredictable in its consequences. The criminalization of landscape burning encouraged the anonymity of burning, further obscuring agency. Although many landscapes across the globe and throughout history have been shaped by fire, not all can be called anthropogenic, even though they may constitute environmental infrastructure to human or non-human users.[70] Finally, despite claims to the contrary throughout recent history, the fact that fire has still not been fully domesticated demonstrates that the interaction of humans with fire should be understood as a process because the outcome is incomplete and insecure.[71]

Fire has a long history in shaping the vegetation of southern Africa and its grasslands and savannas "are particularly well adapted to fire."[72] The landscape of north-central Namibia in the twentieth century was one of the many that directly and indirectly was deeply affected by fire. Fire was not a random or "natural" factor in north-central Namibia even as its semiarid climate and its vegetation "naturally" created the ideal conditions (hot and dry and an abundance of fuel) for fires. Instead, fire occurred within the framework of an indigenous burning regime that was grafted on local environmental conditions. Local inhabitants used fire to create steel tools – hoes and axes – that were critical to the creation and

69 See Goudsblom, *Fire and Civilization*, p. 6. Goudsblom considers the domestication of fire the first truly dramatic ecological transition brought about by humans.

70 See Pyne's trilogy, *Fire in America*; *World Fire;* and *Vestal Fire*. See also O. S. Stewart, *Forgotten Fires*; T. R. Vale (ed.), *Fire, Native Peoples, and the Natural Landscape*; R. Boyd (ed.), *Indians, Fire and the Land in the Pacific Northwest*; Kull, *Isle of Fire*. On the anonymity of burning, see Kull, *Isle of Fire*, pp. 64–66.

71 Goudsblom points out that each generation has to learn how to accommodate to the presence of fire if and when it occurs because fire has been largely eliminated from everyday life. It is the domain of a small group of specialists, and fire control is hardly discussed in any standard textbooks; see Goudsblom, *Fire and Civilization*, pp. 195–215. This perspective is only changing with the discussion about global warming and climate change.

72 Tainton, Zacharias, Hardy, "The Contribution of Veld Diversity to the Agricultural Economy," pp. 107–120 (especially 112–113). See also P. de V. Booysen and N. M. Tainton (eds.), *Ecological Effects of Fire in South African Ecosystems* (Heidelberg: Springer, 1984).

maintenance of environmental infrastructure on and off farm. Compared to stone and wooden tools, metal tools greatly enhanced human's capacity to manipulate the environment.[73] Fire was also critical in transforming vegetation resources into sources of food (including brews and liquor); it provided a warm and light microenvironment around the hearth during cold and dark evenings; and the ashes from fires were spread on the fields, enhancing soil fertility. Moreover, fire was a principal tool to clear bushland to make fields and farms. In addition, people burned the crop residue and weed growth on their fields prior to the onset of the rainy season to prepare the land for the next cropping season.[74]

Colonial officials had no issue with the use of fire as a tool of environmental management in the domestic space of farms and fields. That does not mean, however, that they did not deride its use as primitive. Colonial officials also rued the burning of crop residue and weeds on the fields as an incredible, if short-lived, annual nuisance because it reduced visibility to such an extent that planes were prevented the use of the administration's airstrip at Ondangwa, just as the billowing smoke of annual fires in, e.g., Indonesia and Madagascar hide the landscape.[75] In the context of farmers' preference for old farms and fields, burning fields takes on special significance because given the buildup of (decaying) organic matter, old farms and fields are also very favorable environments for the proliferation of rodents, insects, and weeds. Fire constitutes a powerful tool to control rodents, insects, and weeds, the last especially as compared to labor-intensive hoe weeding.[76] Insects and small mammals are important causes of pre- and post-harvest crop losses as well as sources of diseases, and regular burning creates an unfavorable post-burn microclimate with decreased cover and food, which reduces the undesired populations.[77] Significantly, the long-used Amazonian black earths – the most fertile

[73] Kreike, *Deforestation and Reforestation in Namibia*, pp. 101–123, and Denevan, *Cultivated Landscapes of Native Amazonia*, pp. 27–49.

[74] Kull distinguishes agricultural (on fields), pastoral, slash and burn, and pest control fires that have different timing. See Kull, *Isle of Fire*, pp. 79–115.

[75] On Madagascar, see, Kull, *Isle of Fire*, pp. 15–16.

[76] On fire and weed control, see S. B. Hecht, "Indigenous Soil Management and the Creation of Amazonian Dark Earths: Implications of Kayapo' Practices," and J. Major, A. Ditommaso, L. A. German, and J. M. McCann, "Weed Population Dynamics and Management on Amazonian Dark Earth," in J. Lehmann, D. C. Kern, B. Glaser, and W. I. Woods (eds.), *Amazonian Dark Earths*, pp. 363–364 and 448, respectively. Insect herbivory is more intense in unburned vegetation; see Whelan, *The Ecology of Fire*, p. 192.

[77] On fire and insect and rodent control, see Whelan, *The Ecology of Fire*, p. 192, and Scholes and Walker, *An African Savanna*, pp. 123–124.

of its dark earths – derive much of their fertility and their color from a high level of charcoal and other incompletely combusted organic materials that probably originate from both cooking fires and regular burns. In fact, few fires result in the complete combustion of the above-ground biomass.[78] Susanna Hecht emphasized that "[o]ne of the most surprising discoveries of the Kayapo field work was the realization of the ubiquity of 'in-field burning'" and concluded that the lighter brown soils of Amazonia are associated with regular low-temperature burns.[79] Fires speed up the breakdown of some nutrients and minerals captured in organic compounds, including carbon, but also result in losses, especially of precious nitrogen, which is lost as the greenhouse gas N_2O. In high-temperature fires of more than 600 degrees Celsius, nitrogen loss is virtually 100 percent; at a temperature of 200 degrees Celsius, only half of the nitrogen is lost.[80]

Similar to their colleagues elsewhere, colonial officials in Ovamboland did not appreciate the local importance of the use of fire as a land management tool outside of actual farms and fields. Instead, colonial officials stressed the destructive properties of fire and its uncontrolled use.[81] For example, although Native Commissioner Hahn acknowledged that *inside* the inhabited parts of Ovamboland, fires and the use of fire were strictly controlled by the headmen and kings, he nevertheless dismissed the use of fire as it occurred *outside* the inhabited parts as uncontrolled and devastating.[82]

78 On fire residue in Amazonian dark earths, see B. Glaser, W. Zech, and W. I. Woods, "History, Current Knowledge and Future Perspectives of Geoecological Research Concerning the Origin of Amazonian Anthropogenic Dark Earths (Terra Preta)," in J. Lehmann, D. C. Kern, B. Glaser, and W. I. Woods (eds.), *Amazonian Dark Earths*, pp. 12–13, and E. G. Neves, J. B. Petersen, R. N. Bartone, and C. A. da Silva, "Historical and Socio-Cultural Origins of Amazonian Dark Earths," p. 49. On incomplete combustion, see Whelan, *The Ecology of Fire*, p. 31.

79 See S. B. Hecht, "Indigenous Soil Management and the Creation of Amazonian Dark Earths: Implications of Kayapo' Practices," pp. 362–367. Whelan stresses that fire should not be solely highlighted for its negative impact. See Whelan, *The Ecology of Fire*, pp. 17–20.

80 Scholes and Walker, *An African Savanna*, pp. 81–110.

81 On negative views of burning in colonial and postcolonial Madagascar a country that is considered a textbook case of deforestation as a result of the unbridled use of fire; see Kull, *Isle of Fire*, pp. 49–57.

82 NAN, NAO 45, NCO to Sec. SWA, Ondangwa, December 22, 1938. See also NAO 17, Acting Sec. to Sec. SWA, Windhoek, October [or November] 2, 1928, and NAO 45, Station Commander South West African Police to District Commander Police District Omaruru, Kamanjab, July 29, 1938. See also NAO 10, ANC to NCO, Oshikango, May 30, 1941.

Officials urged the chiefs and headmen of Ovamboland to take strong measures against a wave of clearing bushland for new farms and fields during the 1920s, 1930s, and 1940s, an event that they described in terms of shifting cultivation (i.e., slash and burn).[83] They had both political and economic-environmental objections to the massive and spontaneous squatter movement that marked the 1920s and 1930s. They regarded the new land clearing by those who moved beyond the areas of concentrated settlement in the heartlands of the former Ovambo polities as undermining the authority of the chiefs and headmen who were the pillars of their indirect rule. Furthermore, they considered the massive clearing as a waste of land and forest resources, especially of timber.

Even less controlled and more wasteful did colonial officials judge the use of fire in the "wilderness" areas on the edge and outside of the floodplain. Fires burned and blackened extensive areas to the west of the floodplain (toward Kaokoland), as well as to the south (Okakuejo, Namutoni, Etosha), and to the east, including the Sandveld up to Okavango in the east and Tsumeb and Grootfontein in the southeast. One official described the Sandveld along the modern Angolan-Namibian border during the 1928 "winter" (i.e., during the dry season months of September and October) as hostile to any life because fire had exposed the grey soil and blackened the trees.[84] Another official who traversed eastern Ovamboland further south in late September of the same year wrote:

> I found practically the whole country from Tsintsabis to the Okavango burnt out; fires were raging all over the country as we went up and returned. It is alleged that these fires are started by bushmen. They are undoubtedly doing a great deal of damage and are not confined to the area outside the Police Zone. Fires were raging all over the Grootfontein districts [south of Ovamboland] and are reported to be worse this year than any previous year.[85]

In September 1934, an official traveling in the far western districts of Ovamboland observed "veld fires" in Uukwaluthi, Ongandjera, and Uukwambi. Fires also occurred in the southeastern Ondonga district and

[83] On views of Ovamboland's agriculture as shifting cultivation, see Chapter 4.

[84] NAN, KAB 1, W. Volkmann, October 30, 1928, "Report on the Agricultural and Political Conditions at The Angola Boundary."

[85] NAN, NAO 17, Acting Secretary to Secretary SWA, Windhoek, October [or November] 2, 1928.

intelligence reports indicated that the fires occurring in Ongandjera and western Ondonga originated in Uukwambi. These fires all raged on the edges of the inhabited areas of the districts and the areas between them.[86] The monthly report for August 1936 stated:

> [e]normous tracts of country have been burnt out in the Southern areas of Ondonga. Reports... show that they started in the vicinity of the Etosha Pan and, as a result of the strong winds, spread across the Ombuga flats and the dense bush country lying south of it... large tracts have also been burnt South of Namutoni and near Ekatumare, North of Okaukuejo...[87]

Fires in 1938 were equally devastating. The September report read: "[v]eld fires in Ovamboland, during the dry months have caused enormous damage. In large tracts of country the grazing has been completely burned out. The tribes which suffered most are Ondonga, Ogandjera and U[u]kualuthi." The king of Uukwaluthi sent men out to combat the flames with sticks, "but they found this impossible owing to the enormous front of the fire and driving wind."[88] Only one major veld fire was reported in 1944. But in 1947, "many veld fires" raged across Ovamboland. No major fires were reported during the fire season of 1949, but in 1948, 1950, and 1951, large fires scorched parts of the region.[89] Reports typically attributed such fires to the "wild" bushmen or other "wild" individuals; the fires were thus portrayed as being irrational, wanton, and the result of "playing with fire" without taking precautions. Colonial officials and non-San inhabitants of north-central Namibia alike blamed the "wild" Bushmen, as did the author of the 1928 report on eastern Ovamboland, who stated: "[i]t is alleged that these fires are started by bushmen."[90]

Native Commissioner Hahn believed the Ovambo to be less pyromaniacal than other ethnic groups and reported in 1934 that he had carefully

[86] NAN, NAO 19, Monthly Report Ovamboland, September–October 1934.

[87] NAN, NAO 19, Monthly Report Ovamboland, August 1936.

[88] NAN, NAO 20, Monthly Report Ovamboland, September 1938.

[89] NAN, NAO 20, Monthly Reports Ovamboland, March–July 1941, A450, 7, Annual Report Ovamboland 1943, NAO 21, Quarterly Report Ovamboland, July–September 1943, July–September 1944, NAO 60, Quarterly Report Ovamboland, October–December 1947, April–September 1948, July–September 1949, July–September 1950, and July–September 1951.

[90] NAN, NAO 17, Acting Secretary to Secretary SWA, Windhoek, October [or November] 2, 1928.

explained the administration's fire suppression policy to all the headmen and:

> [t]hey have all undertaken to do their utmost to assist the Administration in this matter. They realise the disastrous effects veld fires would have when the grass and undergrowth is so dense. In previous reports I have mentioned that the Ovambos are not giving [sic: given] to burning the veld as wantonly as is the case with many other tribes.[91]

Those responsible for setting fires were severely punished. After the Assistant Native Commissioner encountered veld fires during an inspection tour through western Ovamboland in September 1934, he asked the headmen of Uukwambi district to investigate and punish the perpetrators. The headmen complied and within a month had fined no fewer than twenty-six of their subjects for arson. But King Martin of Ondonga was less cooperative. In 1936, when enormous fires raged on the edges of and beyond Martin's Ondonga district, the king claimed that he was powerless to check fires that had their origins outside of his territory:

> [h]e [King Martin]...thinks that Bushmen deliberately set the veld alight or that traveling Ovambos were negligent in extinguishing their camp fires. I might mention that at this time of the year the country is so dry and covered with abundant undergrowth and dry grass that a fire, when once it takes hold, is practically impossible to check. It is many years since such large grass fires have been seen in these parts.[92]

Widespread fires in 1938 again spurred investigations into their cause. King Shanika of the Ongandjera district claimed that the fires originated from beyond his territory proper and accused the Herero-Bushmen, who inhabited the area of setting the fires. He forcibly relocated the Herero-Bushmen from their homes to his district in order to better control their actions.[93]

Toward the end of the dry seasons of 1940 and 1941, when the dense grass commenced to dry out, officials traveled through the districts urging the kings and headmen to guard against the occurrence of grass fires, to extinguish any fires that occurred, and to fine heavily anyone who set

91 NAN, NAO 19, Monthly Reports Ovamboland, April–May 1934.

92 NAN, NAO 19, Monthly Report Ovamboland, September–October 1934 and August 1936.

93 NAN, NAO 20, Monthly Report Ovamboland, September 1938.

fires. The war years 1940, 1941, 1943, and 1944 were reportedly fairly free of grass fires.[94] A 1943 report expressed satisfaction that

> [t]he natives are beginning more and more to realise that it is not worth their while to set the veld alight. Heavy fines have been imposed on the wrongdoers in Ondonga as well as in the Ombalantu and U[u]kuambi areas.[95]

In the large Ondonga district where the uncooperative King Martin had been succeeded by King Kambonde, the anti-burn policies of the administration were also being stringently enforced in the mid-1940s. Early in the dry season of 1944, King Kambonde caught three of his subjects "almost redhanded" when they set a fire twenty miles south of Ondangwa on the road to Outjo.[96]

But even as the administration of Ovamboland reported that it was getting wild fires under control, it had become impossible to maintain the fiction that the "wild" bushmen were to blame for the fires. In eastern Ovamboland (the Sandveld) and in the Oshimolo region to the north across the Angolan-Namibian border savanna fires were set intentionally to kill Transvaal teak to facilitate extracting its roots. Gathering the roots was a highly organized business that involved parties of more than twenty men and women from the Ovambo floodplain villages who employed donkeys to transport the roots to the villages. The roots were dried and widely traded across the region because they formed the main ingredient for ochre, which, mixed with animal fat, was used to adorn women's bodies during initiation ceremonies and other festive occasions.[97] A 1943 report stated that "[i]t is generally believed that it is the bushmen who sets [*sic*] the veld alight but it has been definitely proved here that the Ovambos are primarily responsible."[98] In his 1943 annual report, even Native Commissioner Hahn finally conceded that the belief that the Bushmen were the main cause of the fires was erroneous:

> [w]ith the assistance of the more intelligent natives in Ukuanyma [Oukwanyama] and in Ondonga, this office, from information collected, has come to the definite conclusion that the Ovambos themselves are primarily responsible. Bushmen are, of course, not altogether blameless but in this connection it is noteworthy to

[94] NAN, NAO 20–21, Monthly Reports Ovamboland, March–May 1940, March–July 1941, July–September 1943, Quarterly Report Ovamboland, July–September 1944, and A450, 7, Annual Report Ovamboland 1943.

[95] NAN, NAO 21, Quarterly Report Ovamboland, July–September 1943.

[96] NAN, NAO 21, Quarterly Report Ovamboland, July–September 1944.

[97] NAN, SWAA 3, ANC to NCO, Oshikango, December 17, 1941.

[98] NAN, NAO 21, Quarterly Report Ovamboland, July–September 1943.

> repeat the remarks of several of the leading Ukuanyama headmen and also those of chief Kambonde of Ondonga. They state that the veldfires take place in the dry season when the bushmen are more dependent on veldkos, wild fruits, berries, honey, water accumulated in the trunks of hollow trees etc. than at any other time. Why therefore should they destroy this essential source of supply? Further it is persistently alleged that the bushmen set the veld alight so as to bring about an early crop of green grass to concentrate game. If this were the case, the fires would be controlled and confined to smaller areas, whereas at present they are so extensive that they can do nothing else but separate and scatter the game over miles and miles of country making it much more difficult... to hunt, especially if his water supply has been destroyed.... It has been found that in areas such as Western Ukuanyama, Northern Ondonga, Ukaumbi [Uukwambi], Ombalantu, Ukualuthi etc. where there are no bushmen, veld fires occur year after year and as soon as the young grass appears one does not find game feeding there but native stock. This is also the case in many unoccupied regions.[99]

Attributing wild fires to wild Bushmen had not only naturalized fires but also blurred the agency of burning and hid the extent to which the region's inhabitants used fire as a tool to shape and maintain environmental infrastructure. Clearly, Native Commissioner Hahn had a change of heart in the 1940s about the "natural" origins of the annual landscape fires: the fires were not set by wild Bushmen attempting to attract wildlife to facilitate hunting but by Ovambo herdsmen seeking to improve grazing and browsing for their domestic animals.

Whereas reports during the early 1940s stressed a decline in the occurrence of veld fires on the margins and outside of the Ovambo floodplain, reports from the immediate postwar years stress that fires once more were a serious problem. The report for the last quarter of 1947 stated "[that] [m]any veld fires occurred, and the Chiefs and Headmen were instructed to use all their efforts to get their subjects to cease the evil practice of continual veld burning."[100] Fires reappeared during the dry seasons of 1948, 1950, and 1951, although none were reported for 1949. The post–World War II reports seem to contradict the war time reports that veld fires were a thing of the past and that the local headmen were cooperating to prevent them. For example, the 1950 report indicates: "[l]arge areas of North West Ovamboland have been burned off. The evil of continued grass fires has often been represented to the Chiefs and Headmen but they fail to cooperate."[101]

99 NAN, A450, vol. 7, Annual Report Ovamboland 1943.
100 NAN, NAO 60, Quarterly Reports Ovamboland, October–December 1947, April–September 1948, July–September 1949, July–September 1950, July–September 1951.
101 NAN, NAO 60, Quarterly Report Ovamboland, July–September 1950.

It is possible that the early 1940s did see a decline in the occurrence of large fires. Many more men and older boys were absent from Ovamboland during the first half of the 1940s than during the 1930s because of the enormous war-time migrant labor demand in South Africa. Boys and men engaged in herding and hunting in the areas on the margins of the floodplain and beyond, and they were responsible for the use of fire. Moreover, during the early 1940s, because of a deal brokered by Native Commissioner Hahn with his Portuguese counterparts, it was safer and easier to move herds from Ovamboland to dry season pastures north of the border for the duration of the dry season. Burning the veld was often associated with herding and hunting during the dry season, and less intensive use of the fire-prone bushland savanna south of the border might have caused a decline in burning.

But a change in administration early in 1947 was also important in explaining the increased reports of burning after the war years. In order to facilitate his indirect system of rule in Ovamboland, Native Commissioner Hahn had opposed the introduction of many conservation measures that might have affected land use, including, e.g., measures to combat deforestation and to control cattle diseases. Hahn had defended his position by highlighting Ovambo conservation "traditions," including the preservation of fruit trees.[102] This is also the context for the Hahn administration's long-held assertion that the Ovambo did not use veld burning "wantonly" and that the Bushmen were responsible for the destructive annual savanna burning. Hahn's successor, Eedes, who took over as Native Commissioner of Ovamboland early in 1947, however, was little impressed by Ovambo "traditions" and much less reluctant to impose and enforce conservation regulations. Moreover, frustrated in his attempts during his 1947–1953 tenure to eradicate cattle diseases from Ovamboland, he grew increasingly hostile to the Ovambo headmen and to Ovambo practices that he identified as obstacles to progress and development.[103]

Be that as it may, the persistence of bushland savanna burning is also a testimony to the importance of the use of fire as a land management tool; colonial efforts to eradicate the use of fire had still not been fully effective as late as the 1950s. In fact, a disheartened Native Commissioner for Ovamboland proposed more radical measures, including the creation of a fire belt all along the native reserve's southern border to at least protect

[102] See Chapter 2.
[103] Kreike, *Deforestation and Reforestation in Namibia*, pp. 60–73.

white farmlands further south, thereby effectively giving up hope of fully eradicating the use of fire *in* the reserve:

> [m]ost of the grass fires . . . are started by the wild bushmen for the express purpose of attracting game to the green grass which appears about 14 days after the fires. The burning of the grass also enables the bushmen to find certain varieties of veld-kos in an easier manner. Attempts to get the bushpeople to cease the evil practice of continual grass burning have been unsuccessful to date. When the bushpeople are brought under more efficient control, and when steps are taken to improve the lot of these people it may be possible to get them to cease their practice of grass and veld burning. . . . The Ovambo, also on occasion, set fire to the veld. . . . When an investigation is made it is usually reported that the village idiot or a small child started the fire. On occasion the real person responsible is apprehended and is punished by the tribal courts. As reported . . . the Chiefs and Headmen do not co-operate and allow their country to be burnt out on occasion. It seems that more stringent measures will have to be taken against the Ovambos and it is suggested that fines against the tribes be imposed by the Native Commissioners.[104]

The descriptions of fires in the 1920s–1950s reports suggest that they were rarely wild fires in the sense of having natural origins, e.g., as in fires caused by lightning. In fact, most reports of large fires occurred in or before September – that is, before the slow and tortured transition to the rainy season that was marked by lightning storms that could set dry vegetation alight. The fires were "wild," however, in the sense that they were attributed to neglect (carelessness with campfires) or to a primitive or irrational mind (e.g., wild Bushmen, or wandering hunters, careless migrant laborers, or devious herdsmen).

As has been stressed in the rich literature on swidden vegetation in Asia and Latin America, or the emerging literature on the use of fire by Native Americas, Africans, and Australian aboriginals; however, the savanna burning in Ovamboland was not "wild" in the sense of being irrational, random, or unambiguously destructive. The details on the occurrence of savanna fires in Ovamboland from the late 1920s to the early 1950s suggest the existence of a burning regime. The large fires occurred mainly during or before September – that is, toward the end of the dry season. For example, the April–September 1948 report for Ovamboland noted both the local population's expectations that the rains would commence

[104] NAN, NAO 71, NCO, Memo "Grass Fires," August 2, 1951. On attempts to suppress the use of fires, see NAO 26, Assistant Sec. SWA to All Magistrates and Officers, Native Affairs Department, May 8, 1931, and ibid, Administrator SWA, April 16, 1934.

soon, as would the occurrence of fires.[105] Even the largely unsympathetic colonial reports describe the wild fires as grass fires or veld fires that burned dry grass, rather than all-consuming fires that set tree groves ablaze. Reports that highlight that fire is used to destroy individual trees typically refer either to intentionally killing Transvaal teak to obtain the roots as a source of ochre or to clearing fields for the creation of new farms. A 1928 report on the impact of burning along the Angolan-Namibian boundary in eastern Ovamboland refers to bared grey soil and blackened tree trunks in the aftermath of fires, but the trees survived:

> [t]he sandveld beginning at Point 23 is a vast plain with no undulations discernible to the eye, consisting of gray sand often approaching red, overgrown with about 10 m. high sparse forest, consisting of almost only two kinds of trees, under which only high useless grass grows.... The picture of the Sandveld is not an unfriendly one, as long as the trees are green and the game and birds enliven it. In winter, on the contrary when the grass fires have laid bare the gray sand and painted the tree trunks black and no green leaf can be seen and all the game has migrated to the Etosha or the Otjimpolo swamps, it not only appears dead but also inimical to every living thing.[106]

Not only did some colonial sources suggest that fires were neither random nor all-destructive, but research trials in the 1950s and 1960s concluded that the suppression of the use of fire was, in fact, a major cause of bush encroachment. In 1961, after a visit to colonial Namibia the South African deputy secretary of forestry warned that a too intensive level of "forest protection" could lead to a denser understory – that is, it might cause bush encroachment. The deputy secretary of forestry was no doubt very familiar with the results of the research trials. The use of the term "forest protection," probably largely referred to the exclusion of fire from the territories' forests in the three Native Reserves of Ovamboland, Okavango, and Caprivi.[107] Fire was excluded not only from the colony's forest reserves but also from the African reserves and the white settler farms. White settler farmers cut fire lanes along the fences to protect the fences of their farms south of Ovamboland. Thorny tree species, however, rapidly invaded the fire lanes and sometimes even the

[105] NAN, NAO 60, Quarterly Report Ovamboland April–September 1948.

[106] NAN, KAB 1, W. Volkmann, October 30, 1928, "Report on the Agricultural and Political Conditions at the Angola Boundary."

[107] NAN, BAC 131 f. HN 8/17/2, Deputy Secretary of Forestry, "Report of a visit by the Deputy Secretary of Forestry to South West Africa: April 17–29, 1961," Pretoria, May 10, 1961.

adjacent maize fields.[108] Some white farmers noted the correlation between fire suppression and bush encroachment as early as the 1920s. South Africa's Nylsvley research area – a former commercial farm on which fire was suppressed – also suffered from bush encroachment.[109] Colonial fire trials from 1955 to 1961 highlighted that "veld fires" checked bush encroachment. But regular burning was rejected as a measure against bush encroachment because it was considered to be too dangerous and because it killed useful trees. Finally, even though the trials indicated that fires disadvantaged bush more than grass, it was thought to be ineffective in the worst bush encroached areas because grass cover was too light to provide sufficient fuel for an effective burn.[110]

It is difficult to assess exactly the impact of fire on the north-central Namibian landscape as a whole. Iron tools made in local blacksmith's forges, hearth fire ashes, and the annual burning of crop residues and weeds were critical means to create and maintain the environmental infrastructure of fields, farms, and villages. The data are scarcer on the periodic burning of savanna outside the villages, but it is clear that it affected large parts of Ovamboland on a regular basis, some of it perhaps once every two or three years. Other areas may have been little affected by any regular human-initiated fires. Records on veld fires in the Etosha National Park just south of Ovamboland for the 1970s suggest that, on average, 12 percent of the park's surface burned every year and that all vegetation types burned at least once every ten years.[111] The fire-scarred Sandveld that became known as eastern Ovamboland (east of the floodplain) was by the late 1920s frequented by herdsmen on their way to the Oshimolo cattlepost region north of the border and by hunters who had their camps as far east as Eenhana. But the herdsmen and the San and other hunters used the eastern Ovamboland itself much less intensively during that period and human-initiated fires may have been less of a factor. The shaping of eastern Ovamboland's vegetation by fire in the late 1920s created a more open tree-savanna along the border in the western half of the region. The occurrence of very dense elephant bush under occasional dead trees on the reddish soils in the eastern half of eastern Ovamboland in 1928, however, suggests that the impact of fire was far

108 NAN, AGR 545, A. M. Durr to Director of Agriculture, Kalkfeld, June 3, 1952.

109 Scholes and Walker, *An African Savanna*, pp. 46–48.

110 NAN, AGR 563, appendix to Pasture research Officer to DOW Chemical Africa, Windhoek, August 19, 1964.

111 P. E. Stander, T. B. Nott and M. T. Mentis, "Proposed Burning Strategy for a Semi-Arid African Savanna," *African Journal of Ecology*, 31 (1993), pp. 282–289.

from uniform. A denser bush savanna because of a lower fire frequency further east perhaps is also reflected in the game species observed in the different parts. Roan antelope and wildebeest occurred throughout much of eastern Ovamboland in the rainy season, but kudu and elephant were largely confined to the eastern half. Wildebeest is generally more at home in savanna vegetation and kudu and elephant are heavy browsers and are often found in the bushland.[112]

Even in the far east of eastern Ovamboland, however, the impact of fire was highly differentiated, resulting in a mosaic of burned and unburned patches.[113] A 1940 description of the Omaheke in the southeast of eastern Ovamboland depicts a landscape created by fire management, with large trees including the fire resistant Transvaal teak and Rhodesian teak, but with very little undergrowth and abundant grazing that attracted eland. Moses Kakoto was a pioneer settler in the Okongo area deep in the eastern "wilderness." In the early 1950s, he encountered an open savanna landscape at Okongo. Almost all the trees on his wooded plot of land grew after his arrival, except for a few Rhodesian teak trees. He attributed the absence of large trees to the San having used fire to manage the local environment over an extended period:

> [i[t was not like now when people do not start bush fires anymore. [The] San started fires to burn old grass and [so that] new grass would come up.... [Kwanyama] herdsmen did the same as the bushmen and set fire to the bush.[114]

Alpheus Hamundja, who went to the cattleposts in eastern Ovamboland during the 1940s, also recalls that in those days, the veld was set on fire so that fresh grass and leaves would sprout.[115]

[112] On the animals in the late 1920s eastern Ovamboland, see NAN, KAB 1, W. Volkmann, October 30, 1928, "Report on the Agricultural and Political Conditions at The Angola Boundary."

[113] Whelan notes that this is typical in landscapes subjected to regular burns; see Whelan, *The Ecology of Fire*, pp. 152–153.

[114] Moses Kakoto, interview by the author, Okongo (Namibia), February 17, 1993; NAN, NAO 10, ANC to NCO, Oshikango, July 30, 1940, "Report on Development Work undertaken in Eastern Ukuanyama during 1940," pp. 8–9. Amerindian communities in early colonial New England used fire to create a park-like landscape. See Cronon, *Changes in the Land*, pp. 48–51 and 108.

[115] Alpheus Hamundja, interview by the author, Ohamwaala (Namibia), January 26, 1993. In general, it seems, grassland productivity improves in the wake of a fire, and increased flowering and improved seedling establishment may result but the evidence is contradictory; see Whelan, *The Ecology of Fire*, pp. 86–93.

Fire was the most dramatic land management tool employed by the people who used the sparsely populated regions east, south, and west of the floodplain. A burned and blackened landscape rapidly turned green after the first rains because fresh grasses and herbs could proliferate after the old dried vegetation had been destroyed. Fire also accelerated the natural process of recycling nutrients by breaking down dead vegetation. Burned areas not only attracted wildlife by facilitating the growth of fresh sprouts; regular burning also prevented tree cover from growing too dense and out-shading grasses. Finally, burning kept insect and microorganism populations in check, many of them vectors for human and animal diseases.[116] Regular burning may have acted, e.g., as a check on grasshopper and tick populations in north-central Namibia in the 1920s, 1930s, and 1940s. Early in 1925, burning was employed as one of several strategies to destroy locust swarms. Paulus Wanakashimba recalled the event: as a small boy, he helped his family protect their precious millet crop by setting alight heaps of branches and dried leaves to chase away the insects. Reports of grasshoppers in the 1930s suggest that at least some of their local breeding grounds overlapped with the areas that were burned regularly. A December 1934 report noted, "[f]lyers [locusts] are doing damage to crops in Ukuanyama [Oukwanyama district] and natives find it more difficult than usual to deal with this pest because of the scattered nature of the swarms. The uninhabited and dune country surrounding Ovamboland – practically on all sides – is a natural breeding ground." Moreover, 1934 was a year of abundant rainfall with locusts emerging as a threat early during the rainy season and spreading rapidly throughout the year; the same year also saw a high incidence of fires toward the end of the year.[117] In March 1937, large swarms of locusts in hopper stage were reported in the mopane bush fifty miles south of Ondangwa. Large fires were a common occurrence in the same areas. An agricultural officer who compiled a detailed report in 1957 was impressed by the apparent absence of tick borne diseases in Ovamboland; in fact, such diseases do not figure in Ovamboland's colonial reports.[118]

116 Soil sterilization through fire is a little studied topic. Whelan points to a study that demonstrates that five years after a wild fire, the parasite load in a blue grouse population was much reduced. He also notes that in general many insect populations in the litter and soil are heavily reduced after fires, even though the variability is enormous and some recover very quickly. See Whelan, *The Ecology of Fire*, pp. 22, 116, 200–216.

117 NAN, NAO 18–19, Monthly Reports Ovamboland 1925–1940, especially February 1925, December 1934, March–December 1935, and March 1937; Paulus Wanakashimba, interview by the author, Odimbo (Namibia), February 10, 1993.

118 NAN, BAC 133, Agricultural Reports Ovamboland, 1955/1956 and 1956/1957 and BOS, Agricultural Officer Ovamboland to Native Commissioners Ondangwa and

But fire is very selective, suppressing certain species and communities and favoring others. Moreover, fire is a notably imprecise tool to shape landscapes and vegetation. It is clear that once the savanna fires became large blazes they were virtually impossible to control, especially under the scorching conditions that prevailed in September and October when strong dry winds could quickly expand a fire. But the persistent use of the term "grass fire" for even the larger conflagrations, the absence of reports of any upper-story fires that destroy trees, and the presence of fire-induced vegetation (in eastern Ovamboland) suggests the fires were neither as destructive nor as wild as colonial officials sometimes described them. The 1942 Annual Report for Ovamboland demonstrates an awareness of the existence of different types of fire and suggests that the term "grass fire" for the description of the fires in the colonial reports was not randomly chosen. The report notes that a fire which was "essentially a grass fire," burned a large area in the Andoni Flats (south of Ondonga), without trees or other vegetation suffering any real damage.[119] The occurrence of the blackened but live trees among tall grass in the 1928 description of the eastern Ovamboland Sandveld also suggests a lower-temperature "grass" fire.

Regular burning favors herbs and grasses over woody vegetation, but it also favors select woody vegetation species over others, even when it occurs at lower temperatures where grass is the main fuel. Certain woody species thrive under burning regimens; the propagation of Transvaal teak, e.g., is greatly facilitated by the use of fire. The woody fruits of southern African and Australian tree species (e.g., Acacias) protect the seeds from fire in a similar fashion as the northern hemisphere cone trees. Moreover, in the case of most pines and many southern African tree species, fire facilitates and encourages seed release and/or seed flush.[120] Subsoil biomass is typically shielded from the impact of fire and the root

Oshikango, [Ondangwa], August 17, 1956. But many animals use their mobility to escape fires or are protected by their subterraneous shelters. See Whelan, *The Ecology of Fire*, pp. 179, 200–216.

119 Stander, Nott, and Mentis, "Proposed Burning Strategy for a Semi-Arid African Savanna," pp. 282–289; see also *Renewable Resources Journal*, 11 (1993), "Special Report: Workshop on National Parks Fire Policy: Goals, Perceptions, and Reality," especially the contributions by N. Christensen, "Fire Ecology and the Management of Wilderness Ecosystems," and J. C. Billing, "Summary II: Linking Fire and Land Use Policy." See also Nkemi a Tchie and G. C. Gakahu, "Responses of Important Woody Species of Kenya's Rangeland to a Prescribed Burning," *African Journal of Ecology*, 27 (1989), pp. 119–128; NAN, NAO 20, Annual Report Ovamboland 1942.

120 Whelan, *The Ecology of Fire*, pp. 86–97.

system of young Transvaal teak can survive for decades, to re-sprout when the conditions are favorable.[121] Resistance to fire depends on a variety of factors, including bark thickness, the distance seeds and fruits are suspended above the ground, and the density of the undergrowth. In general, however, regular use of fires suppressed trees and bush (because it destroyed young treelings and low bush) and favored grasses, creating an open "park like" landscape. In turn, maintaining open park landscapes under large trees cushioned the impact of fire because it reduced available fuel load; in contrast, an undergrowth of bush enabled a grass fire to jump to the tree canopy.[122] Fire suppression generally favors dense bush and tree cover, which provides more fuel to stoke a fire when it occurs.

Conclusion

Burning and browsing – both managed by the inhabitants of north-central Namibia – created a mosaic landscape that was heavily, but *not* exclusively, anthropogenic.[123] The mosaic landscape constituted environmental infrastructure: it supported humans and their livestock as well as wildlife, e.g. by limiting the buildup of fuel (through browsing and regular burns) and through timed burns to keep fire temperatures low and to increase post-burn biomass productivity and browsing quality.

Browse and burn constituted *environing*. Livestock browsing and burning created and maintained a more open landscape that was favorable to both livestock and wildlife grazers and browsers and less hospitable to such disease vectors as mosquitos, tse tse flies, and ticks.[124] Of course, the result can also be interpreted as deforestation. Preventing access by browsers and suppressing fire leads to a much denser vegetation that – depending on the perspective – can be considered to be

[121] See, e.g., E. N. Sabiiti and R. W. Fan, "Fire Behaviour and the Invasion of Acacia sieberania into Savanna Grassland Openings," *African Journal of Ecology*, 26 (1988), pp. 301–313. On Transvaal teak, see Kreike, "De-Globalization and Deforestation in Colonial Africa," p. 94.

[122] Whelan, *The Ecology of Fire*, pp. 57–134.

[123] The contributions in Boyd, *Indians, Fire and the Land in the Pacific Northwest* stress explicitly differentiate the impact of fire and highlight that many landscapes were little or not at all shaped by fire. Overall, Pyne's trilogy portrays fire and fire regimes in much more homogenizing terms. See Pyne's *Vestal Fire*, *World Fire*, and *Fire in America*. Vale rejects the notion that the Amerindian use of fire shaped the American environment into a cultural landscape; see Vale, *Fire, Native Peoples, and Natural Landscapes*.

[124] See Ford, *The Role of Trypanosomiases in African Ecology*; Kjekhus, *Ecology Control and Economic Development in East African History*; and Schoenbrun, *A Green Place*, pp. 75–76.

positive (reforestation) or negative (bush encroachment). Bush encroachment that occurs when dense bush vegetation suppresses grasses and herbs is considered a serious problem especially in the semiarid regions of Africa, Mediterranean Europe, and southwestern North America, where invasive bush vegetation has rendered large areas unsuitable for commercial ranching. *Environing* through browsing and burning may have been largely responsible for the creation and maintenance of open "park like" landscapes with trees or shrubs.

6

Valuing Environmental Infrastructure and the Myth of Natural Resource Management

Valuing resources, most notably the introduction of land fees during the colonial era, was not a mechanistic outcome of population pressure contributing to land scarcity and "the market" putting a price on the land. Instead, paying land fees reflected the increase of a farm plot's value as a result of investing in and creating and maintaining environmental infrastructure. In north-central Namibia, land fees were only required for turnkey farm plots (i.e., plots that had been cleared and that contained huts, palisades, fences, fields, sources of water, fruit trees, and dark earths). Land fees were not required for plots that lacked any environmental infrastructure.[1]

Reconceptualizing respectively "natural resource managers" and "natural resources" as architects of Nature and environmental infrastructure sheds additional light on natural resource tenure dynamics. Typically, environmental resources are ascribed value because of their real or perceived scarcity, or because of their intrinsic "naturalness," or their religious or nationalistic significance, or their biodiversity or genetic diversity. Rarely, however, are environmental resources that are used or managed by non-Western or premodern individuals, groups, and societies systematically ascribed intrinsic value as a result of the continuous

[1] Schoenbrun identifies critical changes occurring in how resources were valued and controlled in the Great Lakes region of East Africa between 1000 and 1500 AD, including the introduction of land transfers and land fees. The changes coincide with the intensification of settled agriculture requiring heavy labor investments in land clearing. The control, transfer, and "sale" of land (rights), as discussed by Schoenbrun, appear to relate to homesteads, banana groves, and perennial fields (i.e., "environmental infrastructure" rather than land per se); see Schoenbrun, *A Green Place*, pp. 127–216.

investments made by their users.[2] In the colonial and postcolonial eras, a perceived *lack* of investment by indigenous users legitimized massive resource alienation as well as dramatic intervention in local natural resource use in the name of *mise-en-valeur*, economic development, and modernization.[3] Ironically, and more recently, the same argument of a perceived lack of investment by non-Western peoples has been used to reverse natural resource alienation and to support restitution claims by indigenous peoples who are seen to be natural conservationists. Of course, claims and counterclaims about Nature versus Culture, wilderness versus cultural landscapes, scientific versus traditional use, and sustainable versus unsustainable use, underscore that competing environmental conceptualizations and narratives constitute power contestations not only between Westerners and non-Westerners but also between local indigenous groups and within indigenous and endogenous communities. For example, Ovambo farmer-colonists in the 1920s–1950s readily defined the eastern half of north-central Namibia as "wilderness," even though Ovambo and San hunters and herdsmen maintained open landscapes supported by water holes throughout the area, and despite the fact that the very environmental infrastructure that the hunters and herders constructed and maintained facilitated Ovambo farmers' colonization of the area.

North-central Namibian farmers did not perceive themselves to be living either "in" or "by" Nature or merely as users or managers of *natural* resources. Conceptually and physically, they separated their farms, fields, and villages from the surrounding landscape. The homestead constituted the *eumbo/egumbo* and was enclosed by a palisade known as the *omuti*. The farm, its fields, and its fallow were often enclosed by a fence (*ongubu*). The landscapes of farms and fields constituting villages with interstitial browse and coppice bushlands were known as the *oshilongo*, while the uninhabited "wilderness" was known as the *ofuka*. *Ofuka* was where only hunters and herders and their cattle ventured. In contrast, oshilongo was marked not only by the presence of people but also by the infrastructure of farms, fruit trees, crops, food stores, fields, and water and woody resources that made permanent human habitation possible.[4]

[2] Marx held that Nature only gained value through labor; see G. Bankoff and P. Boomgaard (eds.), *A History of Natural Resources in Asia*, p. 3.

[3] For *mise-en-valeur* of forests, e.g., see Guha, *The Unquiet Woods* and Peluso, *Rich Forests, Poor People.*

[4] Kreike, *Re-Creating Eden*, pp. 18–25.

The official colonial view was that the kings or (paramount) chiefs of north-central Namibia owned the land and all the natural resources of any value. Native Commissioner Hahn, who was as influential an administrator as an ethnographer, posited that private landownership in Ovamboland did not exist: "[the] Ovambo have no individual land ownership."[5] In another undated manuscript, Hahn asserted: "[t]he soil belongs to the Chief in his capacity as head of the tribe."[6] In yet another manuscript with the title "Tribal Laws and Customs of the Ovambos," he wrote that in the Ondonga district "[t]he land, people and, nominally all property belong to the Chief."[7]

But converting a piece of bushland into a farm with environmental infrastructure conveyed strong usufruct rights if not absolute property rights to the land clearer. A Church of England (CEM) missionary noted in 1924: "[a]ll property is apparently owned by the chief, but cultivation of the soil gives prior right to the individual that has brought the soil under cultivation."[8]

Paying for Environmental Infrastructure

The introduction of land fees – one-time payments to the local headman upon the allocation of a farm plot – reflects the increase of the plot's value as a result of the improvements that were made to the land by previous occupants. Prime farmland in the floodplain became increasingly scarce in the 1930s, 1940s, and 1950s. Established farms became available only upon the death of the (male) owner or as a result of marital separation, circumstances that often required a long wait, and they came at a high price. In Oukwanyama and Uukwambi, only married couples could apply for land for a farm plot. In a dispute about the allocation of a farm in Uukwambi in 1934, e.g., a (village) headman wrote to the Native Commissioner: "I told the boy that he must first get a woman and that after he was married I would give him a kraal [homestead]."[9] The

[5] NAN, A450, 10, "Agriculture" [Mss.].

[6] NAN, A450, 9 f. 2/38, "Property Rights" [Mss.].

[7] NAN, A450, 9, "Tribal Laws and Customs of the Ovambos" [Mss.].

[8] NAN, NAO 26, Report Ovamboland Cotton Prospects appendix to Alec Crosby to Bishop of Damaraland, St. Mary's Mission, January 11, 1924.

[9] Kreike, *Re-Creating Eden*, chapter 6 and NAN, NAO 100, f. 42/11 (iv), Dadilepo Elia Nakale to NCO, Luderitz, June 15, 1952; NAO 104, file "Ukuambi Affairs 1932–33," Iyambo Nule to Hahn [NCO], Onimnandi, Uukwambi, n.d. [received Ondangwa, June 13, 1934]; NAO 100, I. Risto Kueenda to NCO, Oniipa Opleidingskool, P. K. Ondangwa, October 2, 1952.

couple who received the farm enjoyed lifelong tenure or "occupational rights" of the plot. Upon the death of the owner(s) of the farm, the plot reverted to the (village) headman for reallocation.[10] If the owners moved to another farm plot, they could not take the poles or the roofs from their old home to the new location. That is, they could not compromise or reduce the environmental infrastructure that added value to the plot. Construction materials could be removed only if the homestead site was moved *within* the farm plot.[11] Land fee payments became increasingly common in the 1940s and 1950s because established farms with their improved soil fertility and other infrastructure were highly valued. As a result, adult and married men maintained their residence with their parents for increasingly longer periods of time to save up the sum required to pay the land fee.[12]

The fear of eviction underscored the high value of farms and on-farm environmental infrastructure. Land tenure was not equally secure in all districts and varied in time. Chiefs and headmen, especially in pre-1940s Ondonga, used land eviction as a punishment for crimes and alleged witchcraft activities. Moreover, the individual's inability to "properly" cultivate and maintain a farm frequently served as a pretext for a headman to take away a prime farm and to resell it to a relative, a land-hungry young man, or a wealthy older man. The former occupants were either chased from the village (in the case of crimes or witchcraft) or re-assigned to a much smaller plot.[13] People who were evicted usually

[10] NAN, A450, vol. 23, D6, Land Tenure.

[11] NAN, A450, vol. 23, D6, Land Tenure and vol. 9, "Property rights" and "Chieftainship (Ouhamba)" and vol. 10, "Liability of Ezimo (Epata)" and pencil-written mss with letterhead Suidwes-Afrika-SWA with questions regarding Ovambo law; the first question is: "write about right of appeal"; A450, vol. 8, translation of a letter from Muanyangapu Ambunda to Native Commissioner, dated February 3, 1941. A person evicted from his farm in early 1950s Ondangwa was ordered to remove his property but was explicitly told not to take any of the huts or sticks, NAO 100, statement Nangoro Ninguendja, Ondangwa, October 9, 1952.

[12] In 1954 in Oshamba village in Ondonga, Kapula Shilongo, who already had a tribal trust tax number, still lived on his father's farm, NAN, NAO 91, statement Kapula Shilongo. See also Kreike, *Re-Creating Eden*, chapter 6.

[13] NAN, A450, 9, "Hahn, Rough notes on the tribal customs Ovamboland," pp. 59–60 and A450, 12, SWA Commission: Minutes of Evidence, Sitting at Ukualuthi, August 12, 1935, Hahn, pp. 631–632, and vol. 23, D6, Land Tenure; NAO 9, Administrator [SWA] to Chief Martin, Windhoek, n.d. [May, 1937?] and blue note with heading "Chief Martin," n.d. but inserted between letters dated April 6 and August 26, 1929, and Bjorklund to NCO, Onajena, June 10, 1934; and Enquiry in presence of Council of Headmen, Uukuambi, May 27, 1937; NAO 98, Ushona Shimi to NCO, Ongandjera, October 16, 1947, and NCO to Ushona Shimi, Ondangwa, October 10, 1947, and

were allowed to remain on their farm until after they had harvested any standing crops.[14] A bad omen that Hahn recorded is also suggestive that concerns about eviction ran high during his tenure as Native Commissioner: "[a] view of [a] rat *nembundu*. If seen in day time it is a sign that it is lost and looking for a home [and it] will mean that the man who saw it will be chased out of his kraal."[15]

The high value of land with environmental infrastructure is also illustrated by land strife. Beginning in the late 1920s, in order to consolidate colonial rule and to create a cadre of headmen who would be loyal to his administration, Native Commissioner Hahn elevated a few chosen headmen – who were often his personal favorites – to the newly created rank of "senior" or "councilor" headmen. Hahn purposefully undermined the positions of the former kings by establishing a direct personal and institutional link between the councilor headmen and the local colonial administration. Hahn designated the senior headmen as the heads of subdistricts and grouped them into tribal councils that assisted the colonial officials and the remaining chiefs (kings) in administering the districts. Hahn demoted all the remaining headmen to sub-headmen responsible for a single village.[16] In his efforts to shore up his "senior" headmen in

NCO to Kambonde, Ondangwa, October 10, 1947, Kaimbi Mundjele to NCO, Ombalantu, November 1948 and n.d. (received January 14, 1949) and NCO to Ombalantu Headmen, Ondangwa, November 12 and December 18, 1948; NAO 91, ANC to NCO, Oshikango, October 29, 1953; NAO 104, file "Ukuambi Affairs 1932–33," RCM to NCO, Uukwambi, June 24, 1934; NAO 11, UGR to O/C NAO, Omafo, November 8, 1928, and complaint against Headman Kalipi, July 23, 1928, attached to UGR to O/C NAO, Omafo, November 8, 1928; NAO 100, statement Nangoro Ninduendja, Ondangwa, October 9, 1952; NAO 99, N. Ndjuluua [to NCO?], Ondangwa, December 12, 1950; NAO 105, Diaries NCO, Diary 1928, July 16, 1928; NAO 10, statement Hinikamba, Oshikango, April 1, 1939.

14 NAN, A450, vol. 23, D6, Land Tenure.

15 NAN, A450, 10, "Oshitila."

16 NAN, NAO 10, O/C NAO to NCO, Oshikango, August 31, 1938, and June 26, 1940; ANC to NCO, Oshikango, November 2, 1940; NAO 98, NCO to Kambonde, Ondangwa, October 10, 1947, and Ntinda Shivute to NCO, Oshendje, October 1947 and Kambonde to NCO, Okaroko, October 25, 1947. On the extension of the council system, see A450, 7, Annual Report Ovamboland 1935; NAO 10, O/C NAO to NCO, Oshikango, August 31, 1938, and June 26, 1940; NAO 51, memo NCO, Ondangwa, November 30, 1947, and Allen and Bruwer Blignaut, Proceedings of an enquiry held at Ombalantu on July 15, 1954, into the allegations against Headmen Hishitile Shiweda and Naundjoba Anghuna; NAO 104, file "Ukuambi Affairs 1932–33," Iyambo Nule to Hahn, n.p., n.d., received Ondangwa, November 15, 1934 (two letters), and Iyambo Nule to NCO, Onimnandi, Uukwambi, n.d. [received June 13, 1934]; NAO 51, NCO to CNC, Ondangwa, June 25, 1948, and ANC to NCO, Oshikango, June 16, 1948; Kreike, "Recreating Eden," ch. 4.

Oukwanyama district, Native Commissioner Hahn attempted to grant them monopoly power over clearing land for any new farms: "[i]n Ukuanyama [Oukwanyama] a Native is only allowed to cut down trees when clearing new lands and no land may be cleared without the prior authority of the Councilor Headman of the Area."[17]

Because the cash-starved colonial administration could not pay the headmen for their services in supporting day-to-day rule, the headmen, with the approval and the encouragement of colonial officials, began to demand a one-time land fee when a farm was allocated. They could institute this practice because existing farms had environmental infrastructure and thus were highly valued. Although a land fee had been charged in the form of cattle or other livestock in 1920s Uukwambi, with the amount depending on the size of the plot and the quality of its environmental infrastructure, land fees in Oukwanyama and the Onkolonkathi/Eunda districts were uncommon before the 1930s. By the late 1930s, however, the practice had spread throughout Ovamboland, including the Oukwanyama and Eunda districts, and by 1950, land fees were common in Ondonga district. In Ondonga, Angula paid a land fee of three pounds for his farm in 1927. A decade later, Otto Kasava paid three pounds and fifteen shillings and a "male cattle" for his Eputu farm. In Eunda, the land fee for a farm in the late 1930s to mid-1940s was one head of cattle, while in the Endola area of Oukwanyama the land fee for two farms in 1951 was one ox each. For comparison, during the 1940s, a migrant labor earned one pound per month, an ox cost approximately three pounds, and a steel plow cost more than five pounds. Migrant laborers could only save part of their wages to invest in a farm, but the figures and the data suggest that one or two years of migrant labor were sufficient to earn the land fee. Of course, without further investments in a plow and livestock – as a source of draught power and manure – the farm could not be maintained as a productive unit. Thus, acquiring a plot with environmental infrastructure and maintaining it constituted a considerable and sustained investment that required multiple years of migrant labor to accumulate the necessary capital.[18]

[17] NAN, NAO 44, ANC to NCO, Oshikango, March 24, 1942.

[18] NAN, A450, vol. 23, D6, Land Tenure, vol. 9, "Hahn, Rough notes on the tribal customs Ovamboland," pp. 59–60; and vol. 10, statement Hinikamba, Oshikango, April 1, 1939; NAO 99, N. Ndjuluua [to NCO?], Ondangwa, December 12, 1950; NAO 9, blue note with heading "Chief Martin," n.d. but inserted between letters dated April 6 and August 26, 1929; NAO 99, Otto Kasava to Master Nakale, Namutoni, October 3, 1949, and Ashimbanga Mupolo to NCO, Eunda, October 13, 1950;

Headmen could waive the land fee if a farm was "inherited" by matrilineal (clan) relatives, a practice that was more frequent in the southwestern districts, although it may not have been uncommon throughout colonial Ovamboland. Moreover, a headman could also waive the land fee for applicants who were his relatives. Nevertheless, a land fee typically was required, and the principle of land inheritance was open to challenge by headmen who were under pressure from their own relatives, friends, and clients to allocate farms to them when they became available.[19]

Complaints that the land fees were too high and that they arbitrarily were set by the headmen were frequent. In 1957, a common land fee was one head of cattle. In 1961, the fee varied from 40 to 60 South African Rand (20–30 pounds) plus a head of cattle in Oukwanyama district and 60 to 80 Rand plus one head of cattle or 40 Rand plus two cattle in Ondonga district. In 1967, twenty-five households that lost their farms as a result of road upgrades received monetary compensation based on the size of their holdings. The compensation may reflect land fee levels at the time but probably undervalued the holdings: sixteen households received 1 to 20 Rand each and five received more than 20 Rand. In 1970, the land fee varied from 20 to 200 Rand. Of a 360-household sample surveyed in 1993, 80 percent had paid a land fee, using livestock, cash, or a combination thereof. Cash payments varied from 1 to 1,000 Rand. Most of the respondents specified that the village headman had received the land fee. In 24 percent of the cases, the land fee had been paid to the senior (or councilor) headman.[20]

NAO 100, Dadilepo Nakale to NCO, Luderitz, June 15, 1952; Kreike, *Re-Creating Eden*, chapter 6; Kreike, *Deforestation*, pp. 134–135.

[19] NAN, NAO 10, statement Hinikamba, Oshikango, April 1, 1939; NAO 55, Roads Department to NCO, Windhoek, May 14, 1952, and ANC to NCO, Oshikango, June 14, 1952; NAO 104, file "Ukuambi Affairs 1932–33," RCM to NCO, Uukwambi, June 24, 1934; NAO 55, NCO to ANC, Ondangwa May 6, 1952, translation of a telegram from Spota Kakelo, Okahandja, May 6, 1952, and telegram [?] NCO to Spota Kakelo, [Ondangwa], May 8, 1952; NAO 99, NCO to Chief Kambonde, Ondangwa, May 6, 1950; Augustus David to NCO, Abenab Mine, April 29, 1950; and NCO to Magistrate Grootfontein, Ondangwa, May 29, 1950. On relatives and land fees, see Paulus Wanakashimba, interview with the author, Odimbo (Namibia), February 10–11, 1993. For an example, see NAN, NAO 91, statement Titus Muatelai Kakonda, Ondangwa, May 10, 1954.

[20] NAN, BAC 133, Agricultural Report Ovamboland 1955/1956; BAC 45, Minutes Tribal Meeting Oukwanyama, June 12–July 24, 1961; BOS, NA to Omuhona [T. G. Strydom], H 15, North Camp, Oranjemund, March 27, 1961, and Tribal Secretary G. Kautwima to Omutonateli Wowilonga, Ohangewena, March 1, 1967; OVJ 15, "Minutes of the elected Committee on land ownership..." Oshakati, December 4, 1970, Appendix Secretary of the Interior to Secretary Justice, Ondangwa, November 9, 1973; OVA 53,

As the value of farm and field environmental infrastructure increased in the 1940s and 1950s, single women with good farms became a principal target of greedy village headmen. In fact, most of the documented evictions, which often resulted in the subsequent subdivision of the farm plot and its resale, involved single (typically elderly) women or women whose husbands were away on long migrant labor contracts.[21]

Once a pioneering village with rudimentary environmental infrastructure became well endowed with farms, fruit trees, water sources, fences, and fertile fields, it was often subdivided. The practice of sub-dividing villages, which could contain 10 to 100 farms each, mushroomed in the 1940s and 1950s. In 1938, Holongo Ameshelelonanda paid a fee of 10 pounds (20 Rand), two oxen, and two heifers to King Martin of Ondonga for the position of village headman for a section of Oshiyagaya village after the village had been subdivided.[22] Often, the colonial administration was unaware that the villages had been subdivided. One reason for the secrecy may have been that officials wanted to limit the number of households per village to prevent "overpopulation." In 1956, the agricultural officer for Ovamboland reported that chiefs and headmen derived "great richness" from the sale of farms "and would therefore not support any scheme that establishes the number of kraals per ward [village] based on the agricultural potential of the said ward."[23] In 1952, an investigation concluded that Ondonga had more than 600 villages but the colonial administration had no record of 25 percent of them. The rapid increase in the number of villages in Oukwanyama in the 1950s and early 1960s was partly a result of the founding of new villages in sparsely inhabited areas (principally in eastern Ovamboland), and partly because of the subdivision of old villages.[24]

Sec. SWA to Sec. Agriculture Owambo, Windhoek, June 24, 1974, Appendices A–C; OMITI 5.1.2–5.1.5.

21 See, e.g., NAN, NAO 98, Dalengelue Aitana to NCO, Ombalantu, January 21, 1948, and NCO to headman Dalengelue Aitana, Ondangwa, January 26, 1948.

22 NAN, NAO 100, Chief Kambonde to NCO, August 8, 1952, and Statement Holongo Amshelelonanda at Ondangwa, July 4, 1952; NAO 104, file "Ukuambi Affairs 1932–33," Iyambo Nule to Hahn, n.p., n.d., received Ondangwa, November 15, 1934 (two letters) and Onimnandi, Uukwambi, n.d. [received Ondangwa, June 13, 1934]. See also NAO 99, statement Mingana Shikongo, October 28, 1950, and NAO 11, NCO to Clarke, Ondangwa, November 12, 1936.

23 NAN, BAC 133, Agricultural Report Ovamboland, 1955/1956.

24 NAN, NAO 51, NCO to Chief Kambonde, Ondangwa, December 12, 1951, and Chief Kambonde to NCO, Okaroko, April 18, 1952; NAO 100, Chief Kambonde, August 8, 1952, to NCO and Statement Holongo Amshelelonanda at Ondangwa, July 4, 1952; Chief Kambonde to NCO, Okaloko, January 21, 1954, and NCO to Chief Kambonde,

The headmen who had been selected by the colonial administration as senior or councilor headmen in the 1930s raised revenue by selling villages after a village headman had passed away or had stepped down. Because the villages and subdistricts contained a finite number of farms and villages respectively, and because the farms and villages were allocated for a lifetime, a quick way to increase revenue was to subdivide the farms and villages. The practice was attractive not only to the headmen because it allowed them to raise revenues, but it was also very appealing to buyers because they acquired a plot that either already had environmental infrastructure (a fence, fields, coppice bush, fruit trees) or it was within easy reach (neighbors' water holes, neighborhood or village wells, water reservoirs, coppices, forages, and pastures), as opposed to having to clear a farm from scratch in the uninhabited areas. Alternatively, village headmen could allow a few more applicants to clear a farm in an old village. In the 1950s, land in the floodplain had become so precious that at times headmen even succeeded in extracting a land fee for the uncleared bush plots in or near their villages because applicants could rely on the environmental infrastructure of neighbors and the village as a whole.[25]

Valuing and Owning Trees

Even as colonial officials maintained that the abundant fruit trees were wild and wilderness trees, they were very much aware of the extent to which trees were valued and their ownership contested. Native Commissioner Hahn noted that "[i]n fact certain species are very jealously guarded and many species have special names and there are definite laws in regard to their ownership."[26] Indeed, Paulus Wanakashimba recalled that fruit trees were not owned by "the household" or by "the head of the household," but by the owners of the individual fields: "a palm tree belongs to the owner of the particular piece of land. If it is in a woman's piece of land then it is the wife's palm tree. All trees within a garden

January 19, 1954; BAC 44, Meeting held at Ombalantu, September 19, 1960; AHE (BAC) 1/2, Bantu Affairs Commissioner Oshikango to Chief Bantu Affairs Commissioner Windhoek, Oshikango, December 17, 1964, and Minutes of a Meeting held at Okalongo on January 11, 1965.

25 NAN, BAC 44, Minutes of Meetings, June 7–21, 1957; BOS, NCO to CNC, Oshikango, June 24, 1958; NAO 10, statement Hinikamba, Oshikango, April 1, 1939.

26 NAN, NAO 44, Officer Oshikango to NCO, Oshikango, March 17, 1931, and NCO to Secretary SWA, Ondangwa, April 20, 1931.

when they are taken care of – if in a woman's garden then [they] are her birdplum trees."[27] Petrus Mbubi in 1993 asserted that his parents had owned all of the thirty-nine fruit trees that grew on their farm in Oikokola (Ondonga) during the 1920s.[28]

Hahn was baffled by what seemed to be contradictions in who owned which trees when different parties in a dispute confronted him with different versions of "traditional tenure" and consequently attempted to codify local resource tenure in the context of consolidating colonial rule. His main sources were missionaries, selected chiefs and headmen, and his Ovambo assistants and servants. One of his sources, probably a Finnish missionary stationed in Ondonga, noted that "[i]n some districts the fruit trees in the field do not follow the ownership of field, but must be purchased separately."[29]

Not surprisingly, given his preferred sources, Hahn's codification of customary Ovambo law tended to favor claims by chiefs and headmen. His rough personal notes, based on information he gathered during the 1920s, attest to the enormous variation in tree tenure. In Ongandjera, "[the] [o]ccupier [of a farm] has not always got the right to all the fruit trees in area allotted him as in many areas fruit trees are reserved for chiefs (fruit for beer making) and occupiers understand this condition." In Onkolonkathi and Eunda on the southwestern edge of the floodplain, "[f]ruit trees are general(ly) shared [in] particular marala [marula] tree by the headman who has right to collect fruit from certain trees reserved for himself. Owner of kraal may not use fruit from reserved trees unless he receives headmans permission." In Ombalantu, Hahn was persuaded that the headmen's control over land and trees had collapsed.[30] During the early 1930s, Hahn concluded that in Oukwanyama district:

> Omuongo [Omwongo] (marula) trees and fruit belong to the chief [king] but the headman of the omukunda [village] owns the fruit which remains after the chief had his share collected; it is the duty of the headmen to collect it and to send it to the Chief. If a chief does not require the fruit from a particular area the headman owns it. Today, in the absence of a chief, the eight principal headmen exercise the rights of the chief in this respect; each in his own district.... The right of use of indigenous fruit trees other than omuongo... depends upon the situation of

[27] Paulus Wanakashimba, interview by the author, Odimbo (Namibia), February 10–11, 1993. On the *oshikokola-plot*, see Kreike, "Recreating Eden," ch. 7.

[28] Petrus Mbubi, interview by the author, Onanime (Namibia), February 26, 1993.

[29] Kreike, *Re-Creating Eden*. On the missionary, see NAN, A450, vol. 9, "Ovambo Customs (1926)."

[30] NAN, A450, vol. 23, D6, Land Tenure.

such trees. If a tree stands in a cornfield the fruit belongs to the cornfield and the possessor or occupier for the time being becomes the owner of such fruit. If the tree is outside the cornfield or kraal its fruit is public property.[31]

Headmen's claims, however, were contested. In 1939, e.g., a farmer from Uukwaluthi district in the far west appealed to the Assistant Native Commissioner at Oshikango to foil the local headman's attempt to take control over his marula tree. In 1947, a farmer from Ondonga district refused to give his headman a portion of the marula wine that he had produced with fruit harvested from his trees.[32] Native Commissioners Hahn and Eedes assumed the rights of the kings to have passed on to their administrations. Consequently, in a 1953 letter to Petrus Kuferi, Eedes, wrote: "I notice that you are concerned about what you call YOUR marula trees [emphasis original]. The trees are not yours – they belong to the Government. You are really worried about them because they produce a form of liquor."[33]

A broad interpretation of the concept of the "moral economy" sheds light not only on the contested claims to fruit tree ownership but also on some of the contradictions.[34] Gifts of fruit and fruit products assured chiefly patronage and reciprocal access to fruit trees between neighboring households, permitting the residents of marginalized households (especially elderly single women) access to either "private" or "communal" fruit trees. For example, Shangeshapwako Hauladi's family lived in Onelombo (in modern Angola) until the late 1920s. Her family had no fruit trees on their farm but jackalberry and fig were abundant throughout the village and "anybody could come and gather the fruit from any tree." In Omundaunghilo in eastern Ovamboland in the 1920s, birdplum trees similarly abounded and every household had fruit trees. Children could go anywhere to pick the birdplum fruit. Paulus Nadenga's parents' neighbors gathered fruit from his parents' birdplum and marula trees; the neighbors in turn allowed his family to gather fruit from their marula tree, which bore fruit later in the season.[35]

31 NAN, A450, vol. 9, "Property Rights."

32 NAN, NAO 10, Statement Hinikamba (Onkolonkathi), April 1, 1939; NAO 98, Kambonde to NCO, Okaroko, November 4, 1947.

33 NAN, NAO 63, NCO to Petrus Kuferi, Ondangwa, December 10, 1953.

34 On the concept of "moral economy," see J. Scott, *The Moral Economy of the Peasant: Rebellion and Subsistence in Southeast Asia* (New Haven: Yale University Press, 1976).

35 Interviews by the author: Shangeshapwako Hauladi, Omundaunghilo (Namibia), July 14, 1993; Paulus Nandenga, Oshomukwiyu (Namibia), April 28, 1993; see also Helena

The contestation concerning the ownership of fruit trees on private farms and the rights to their fruits that emerged during the early 1970s underscores the value of tree fruit and its products. A committee of the new Ovambo "homeland's" legislative council undertook to evaluate land tenure in Ovamboland, holding closed meetings with a limited number of select notables in all the "tribal areas." One of the objections raised against existing land tenure arrangements was that if a person "bought" a piece of land with fruit trees, the fruit would not belong to the buyer. The committee concluded that "people" demanded that "[f]ruit trees belong to the owner of the parcel of land and if a headman wanted, e.g., marula drink, he should receive this as a percentage and indeed in a friendly way."[36]

In the 1970s and 1980s, the king and headman in the southern Ondonga district allegedly still owned the fruit trees (including the marula), even if they were located on an individual's farm. Commenting on tree management practices, 18 percent of the 1993 OMITI respondents emphasized that in the past, especially in the case of (marula) fruit trees, outsiders (specifically headmen and kings) had rights to fruit trees on a household's farm. Individuals who picked the fruits solely for their own use were punished with heavy fines. In Oukwanyama district, however, where royal authority had been destroyed during the conquest of the 1910s, the owner of a farm had full use rights for all of the fruit trees on the land except for the marula tree: part of the marula wine had to be presented to the village headman, who in turn gave a share of it to his district headman (or senior headman). For other fruit trees, however, the owner of the farm could even suballocate use rights to a third party.[37]

By 1993, it was rare that outsiders had rights to the fruits of a household's on-farm trees. When asked if they would change any rules regarding trees, just 6 percent of households answered that the owner of a farm

Nailonga, Ekoka (Namibia), February 23, 1993, and Mateus Nangobe, Omupanda (Namibia), May 24, 1993.

36 NAN, OVJ 15, "Minutes of the elected Committee on land ownership and use," Oshakati, December 4, 1970, Appendix to Secretary of the Interior to Secretary Justice and Labor, Ondangwa, November 9, 1973, and "Kommentaar op konsepverslag van gekose komitee van die Owambo wetgevende raad . . . [1974?]."

37 NAN, BOS, N. A. to [Bantu Commissioner Ondangwa], Oranjemund, March 27, 1961, and Native Commissioner Oshikango to CNC, Oshikango, June 24, 1958, and October 8, 1961. Fruit trees and bundled rights, OMITI 4.1.8. Cf. OMITI 4.4.37 in which 41% of a small sample of the respondents (*N*=98) emphasized that in the past households did not enjoy exclusive rights to the (fruit) trees on their farm because of rights held by kings, headmen, church elders, and the individuals who had planted the trees involved.

should fully control the on-farm trees. According to 79 percent of OMITI respondents, the head of the household, and/or the members of the household (according to 67 percent of the respondents) could freely gather fruit from their on-farm fruit trees. Only 2 percent of the respondents stated that anybody could gather the fruit of on-farm trees and a mere 1 percent said that fruit from trees could be freely picked from a neighbor's farm.[38]

In sharp contrast, according to 84 percent of the respondents, the fruit of any off-farm trees could be gathered freely by anybody. Off-farm trees were sometimes fenced, an action that constituted a general right according to 51 percent of the respondents, although 27 percent of the respondents maintained that no one was allowed to fence off-farm trees.[39] Over time, local understanding surrounding on-farm versus off-farm fruit tree rights clearly evolved. Valuing trees in terms of investments in environmental infrastructuring was not limited to north-central Namibia, but is common throughout Africa and often involves individual ownership claims, as is the case e.g., with Shea butter in West Africa.[40]

Valuing and Owning Water

The expansion of the colonial water infrastructure in the 1950s, 1960s, and 1970s, had significant effects on water access and management. "Traditional" water holes were overwhelmingly "private" resources controlled by individuals and individual households.[41] Repair, maintenance, and water hole construction relied on locally available tools, materials, and knowledge. Under colonial rule, some of the best sites for harvesting water through water holes were transformed into water storage dam facilities under vague notions of "communal" tenure that on occasion resulted in de facto alienation of household and village water infrastructure. In the 1950s and 1960s, e.g., the administration requested

[38] OMITI 4.1.9 and 4.1.2.0–1.

[39] OMITI 4.1.2.0–1 and 4.1.6.1.

[40] D. R. Harris, "Traditional Systems of Plant Food Production and the Origins of Agriculture in West Africa," in J. R. Harlan, J. M. J. de West, and A. B. L. Stemler (eds.), *Origins of African Plant Domestication*, pp. 311–356 (especially 326).

[41] On the private ownership of water holes, see, e.g., NAN, WWA [WAT?] 637, Department Water Affairs, Report Re. Watersupply Ovamboland Schools, Appendix to Acting Chief Inspector to Director Water Affairs Windhoek, Otjiwarongo, May 13, 1970.

that African owners of individual water holes and wells allow colonial dams to be constructed on their sites because the dams would benefit the community as a whole. Memories of this sacrifice by local inhabitants, however, were short-lived. In 1970, the management of Onandjokwe hospital prohibited local villagers' access to "its" dam even though Johannes Nkana claimed that his father had owned a water hole on the site where the dam had been constructed in 1958.[42] In 1970, the doctor at Onakayale hospital requested that the colonial administration construct a second dam at Onakayale village so that the existing dam could be reserved for use by the hospital, the labor recruitment depot, and the local school only. The doctor explained that although most of the local households relied on water holes, during drought conditions they had come to depend on the water taps at the hospital and the school that drew on the dam reservoir.[43]

Colonial officials typically considered the new colonial water storage dams to be government property. In addition to highly specialized technology and materials, constructing a dam (as well as repairing and maintaining it) required mobilizing a large labor force or a bulldozer. Individual households and even an entire village did not have the capacity to repair and maintain a colonial water storage dam, let alone to construct one. Dam ownership was vaguely defined, sometimes as "communal," but more often as "governmental," which frequently was personified in the form of the local representative of the colonial government: the district headman.[44]

Although headmen were held responsible for maintaining the dams, colonial officials interfered frequently. In 1957, e.g., the colonial administration decided to equip all new colonial dams with goat-proof fencing. In addition, the administration installed hand pumps to limit the number of cattle that could be watered at each dam, with the aim of preventing overgrazing and trampling.[45] Finally, in the mid-1960s, the

[42] NAN, WAT 148, Director Water Affairs to Bantu Commissioner, Windhoek, March 23 and 24, 1966, and BOS, District Record Book Oshikango, pp. 10–11; OVA 33, Senior Magistrate Ondangwa to Superintendent Onandjokwe Lutheran Hospital, Ondangwa, March 4, 1970.

[43] NAN, WWA 649, Dr. Enzlin to Director Water Affairs, Onakayale Hospital, January 18, 1970.

[44] NAN, WWA 637, Report Watersupply Ovamboland Schools, Appendix to Acting Chief Inspector to Director Water Affairs Windhoek, Otjiwarongo, May 13, 1970.

[45] NAN, BOS, District Record Book Oshikango, pp. 10–11. BAC 44, Minutes quarterly meetings held at four centers in the Oukwanyama Tribal area, June 7–21, 1957.

PHOTO 6. Fenced off-farm marula tree (photo by the author, 1992).

colonial administration prohibited wading, swimming, and fishing in the dams.[46]

By the early 1970s, agricultural extension officers effectively supervised dam maintenance and repair. They inspected the dams, reported damaged fences, and held meetings with the local populations to urge them to maintain the dams and to encourage them to dig new ones. Coincidentally, people increasingly began to expect the government to care for existing dams and to build new ones. During the second half of the 1960s, e.g., only one senior headman, Vilho Weyulu of the Oukwanyama district, was praised for his careful water management.[47] The water storage dams were fewer in number and therefore less readily

[46] NAN, WAT 148, Director Water Affairs, Circular to all Departments and Branches of the Administration of Ovamboland, Windhoek, January 4, 1966.

[47] Agricultural extension officers' reports, NAN, OVA 61 and OVA 59. See also OVA 33, letter initialed S. L. B., n.p., January 18, [1973] and memo Oukwanyama Tribal Authority to Director Works, Ohangwena, January 11, 1973; Headman Mupupu Ngwali to Director Works, Onankali, September 25, 1972; Director of Works Ovamboland to Director Works, Ondangwa, December 4, 1971. On Weyulu, see AHE/BAC 1/2, Questionnaire on Vilho Weyulu 11/3/1968.

available to households than the numerous water holes. An agricultural extension worker who in 1971 urged village headmen and villagers to fence their dams was asked, "how can you come here to talk about dams; we have no dams and pumps here. The dams are far away. But we would like the department to dig dams for us."[48] The expansion of a colonial water infrastructure, even as it cannibalized indigenous water infrastructure, did not entirely replace the latter. In 1991, 60 percent of households in north-central Namibia relied on water holes and wells for their daily needs.[49]

The Myth of *Natural* Resources Management

Environmental infrastructure was highly valued in north-central Namibia from the 1930s to the 1990s. The high land prices in the period were not a reflection of any land scarcity per se: land without environmental infrastructure was abundant and available for free. Instead, only land with environmental infrastructure required payment of a land fee. Farms and villages were the result of a substantial and continuous investment of labor, capital (saved through migrant labor wages), and other resources in the land. The heavy investments provided the engine and fuel for both physical and mental processes of environing: households cleared land and constructed homesteads, fields, and water holes and reimagined insecure *ofuka*-wilderness into sheltered *oshilongo*, a landscape filled with farms and fields, coppice bushland, and pastures. The high value of farms and villages that boasted environmental infrastructure found expression in the substantial transfer fees that village headmen and chiefs respectively could demand from applicants to farms and village headmanships. The value of a village – as reflected in its transfer fee – was directly derived from the combined values of the farm infrastructure in the village concerned.

The farm owners who had invested in the environing, however, neither sold the farms nor received the land fees. Instead, village headmen "sold" farm environmental infrastructure because the colonial administration delegated the right to them. The Native Commissioners of Ovamboland regarded themselves as the legal successors of the Ovambo kings who they

48 NAN, OVA 61, Monthly Reports Agriculture, Moses Nandjebo, report for August 1971.

49 Namibia, 1991 Census, Report A, Statistical Tables, vol. v, table H06. OMITI 3.15, 3.16, 3.17, 3.18, 3.19. For private water holes, see also, e.g., NAN, WWA [WAT] 637 f. 31/3/1(ii), Department Water Affairs, Report Re. Watersupply Ovamboland Schools, Appendix to Acting Chief Inspector to Director Water Affairs Windhoek, Otjiwarongo, May 13, 1970.

considered to be the owners of the land. Following the colonial logic, technically, the land fees should have accrued to the colonial administration, but the administration allowed the headmen to receive the fees in lieu of paying a salary, solidifying the system of indirect rule that marked colonial Ovamboland.

Land fees in Ovamboland and elsewhere did not constitute the purchase prices for the land per se. Instead, a land fee represented a one-time lifetime usufruct rent paid by the user or – in the case of north-central Namibia between 1930 and 1990 – a rent for using the environmental infrastructure on the land. It cannot be overemphasized that land fees were not levied on land that lacked environmental infrastructure. With the end of colonial rule in Namibia, however, land fees have become less common. At the same time, the ability to inherit land that contains environmental infrastructure has become more common. The trend predated the end of colonial rule, but farm inheritance has been institutionalized and formalized after independence. Inheriting farms was on the rise before 1990 and inheriting trees, water holes, and wells goes back at least to the early colonial era. Inheriting farmland during the heyday of colonial rule was somewhat more difficult to reconcile with the colonial claim that the administration owned the land. Inheriting water holes was widely accepted locally because the colonial administration paid little attention to them as a resource, whereas the inheritance of wells was not an issue to the colonial administration because of the obvious heavy investments required for their construction and the fact that they mostly had been initially constructed by missions and prominent headmen. Like water holes, fruit trees initially also remained under the colonial radar screen and locally their ownership and inheritance was generally accepted, although from the 1950s onward they became a source of contestation. Thus, although fruit trees, water holes, and wells were not bought and sold, they were very highly valued as environmental infrastructure and as a result were considered to be the inheritable property of the individual(s) who protected, planted, or created and managed them.

Valuing the process of creating and maintaining environmental infrastructure can be read from the payment of land fees in Ovamboland and elsewhere, but also is inscribed in the practice of acknowledging the first known settlers in an environment – that is, acknowledging the creators of the "original" environmental infrastructure as the "owners" of the land. The designation frequently not only is exclusively symbolic, but is also a reflection of the fact that the environmental infrastructure created and maintained by the firstcomers – e.g., the San and Ovambo hunters and herdsmen in eastern Ovamboland – was critical for subsequent settlers. This phenomenon is highly prevalent throughout Africa and beyond. For example, on Madagascar, most permanently cropped fields are considered to be "privately" owned. In the case of fallow lands, the descendants of those who originally cleared the land are considered the owners.[50] Boomgaard argues that in South East Asia, wet rice lands – that is, the arable that had been created and maintained through environing – was owned by individual households and not the ruling aristocracy.[51] In fact, Javanese and Balinese inscriptions from the ninth through the fifteenth centuries highlight that farming practices included irrigated rice fields, rainfed fields, swidden and swidden fallow, home gardens, orchards, betel groves, irrigated rice, taro fields, and cotton fields. Often, the rice fields were individually owned, inheritable, and alienable. Three types of land were sold and bought: home gardens (with trees and perennials), orchards, and irrigated rice fields. The last were valued highest in sale and tax terms. The ownership and sale value of home gardens, orchards, and irrigated rice fields was a reflection of the heavy investments that the owners made in creating and maintaining environmental infrastructure.[52] In South Asia, land tenants who derived land rights from claims that their ancestors had brought the land under cultivation, constituted a separate class of farmers.[53] In Late Imperial China, conflicts between landlords and tenants and the development of the "two-lords-to-one-field" system can be read as an acknowledgment of the increased value of the landlord's property as a result of their tenants' environing investments. As landlords decreased the inputs they provided to develop the land (seeds, tools, animals), tenants increased their investment by constructing and maintaining environmental infrastructure. Tenants subsequently began to claim ownership rights, especially when landlords attempted to "resell" the land. A tenant and a landlord sharing ownership of a field through the "two-lords-to-one-field" system provided more security for the tenant "and offered him incentives to

[50] Kull, *Isle of Fire*, p. 32.

[51] P. Boomgaard, *Southeast Asia: An Environmental History* (Santa Barbara, CA: ABC Clio, 2007), p. 71.

[52] J. W. Christie, "Water and Rice in Early Java and Bali," in P. Boomgaard (ed.), *A World of Water*, pp. 235–258 (especially 243). Planted trees in West Kalimantan were owned by the planters, were often named after the planters, and were inherited by the children of the planters. Moreover, inheritance of a tree did not in and by itself convey full "ownership;" labor investment in its maintenance and use was as important in conveying rights to trees, see Peluso, "Fruit Trees and Family Trees in an Anthropogenic Forest," pp. 54–102.

[53] D. Ludden, *The New Cambridge History of India: IV An Agrarian History of South Asia* (Cambridge: Cambridge University Press, 1999), p. 212.

improve the soil." In eighteenth-century Hunan, tenants paid a one-time deposit to the landlord upon being allocated land, a system very much akin to the land fees paid to the chiefs and headmen in twentieth-century Ovamboland.[54]

The idea of valuing environmental infrastructure and environing is not confined to the "non-West," as attested to by Locke's idea that labor investment created (private) property and Marx's elaboration that Nature gains value through (human physical) labor, thus creating a "second nature" of artifice.[55] Still, although the argument regarding environing elaborated previously stresses that the creation, maintenance, and reproduction of environmental infrastructure involved significant investments of labor, capital, and other resources, it also introduces a twist to the Lockean and Marxist theories. Environing involves *multiple* human *and* nonhuman agency and the object and subject of environing are blurred: humans do not act upon passive Nature or Nature upon passive humans.[56] Instead, both are implicated as actors and are acted upon. Therefore, environing as such does not result in exclusive property rights as per the Lockean principle: multiple human actors may be involved as well as multiple nonhuman actors and forces.

Reconceptualizing "land" as environmental infrastructure also sheds new light on a supposed African exceptionalism: Africa has been treated as having unique historical dynamics because of the presumption that land was abundant while human labor was in short supply. In precolonial Africa, so the argument goes, conflicts were not about land and territory but about control over people.[57] The same argument has equally

54 Perdue, *Exhausting the Earth*, pp. 145–163.

55 On Locke, see F. von Benda-Beckmann, "Contestations over a Life-Giving Force: Water Rights and Conflicts, with Special Reference to Indonesia," in P. Boomgaard (ed.), *A World of Water*, pp. 259–277, and N. Graham, *Lawscape: Property, Environment, Law* (Abington, UK: Routledge, 2011), pp. 45–50. On Marx and natural resources, see Worster, *Rivers of Empire*, p. 26; G. Bankoff and P. Boomgaard (eds.), *A History of Natural Resources in Asia: The Wealth of Nature* (New York: Palgrave MacMillan, 2007), p. 3; K. Lange, "Energy and Environmental Security: The Syr Darya Crisis of Central Asia," in R. Coopey and T. Tvedt (eds.), *A History of Water, Volume 2*, pp. 404–429 (especially 406); and Gadgil and Guha, *This Fissured Land*, pp. 12–13.

56 Graham emphasizes that land as property continues to be defined within the confines of a Nature–Culture dichotomy because the Lockean principle of human agency creating property by working passive Nature is still dominant. She argues that both the fields of property law and environmental law operate within a Nature–Culture dichotomy. See Graham, *Lawscape*, especially pp. 160–173.

57 Scarborough not only reiterates the low population/abundant land premise for Africa but also implies (in Boserupian fashion) that with the absence of pressure on the land, there

been advanced in other regions with low population pressure – e.g., the Amazon, Central America, and parts of India and Southeast Asia, nearly always in the context of "tribal" peoples who engage in hunting and gathering or shifting cultivation.[58]

Because land was abundant, elites had to bind people to them by other means. One strategy was violence: enslave people or exact tribute. A second was to use kinship (marriages and emphasizing lineage or clan relationships), including the creation of artificial kinship. A third strategy was to create patronage networks through the redistribution of scarce goods, including slaves, women, and children, as well as such European imports as guns, horses, textiles, and liquor.[59]

In this context, the African slave trade – exporting an already scarce resource – constitutes a puzzle, leading Walter Rodney e.g., to argue that the export of people (labor) underdeveloped Africa. If Africa already was underpopulated and short of labor on the eve of the slave trade, the labor losses caused by the slave trade, including not only captives who boarded the slave ships but also the people who perished or were maimed and traumatized by the violent slave raiding, must have been devastating. At the very least, the argument helps explain why Africa as a continent remained underpopulated until the mid-twentieth century.

Because of the slave trade or not, Africa's vast land mass and relatively small populations may make land appear abundant in Africa. Most of Africa's land is not, however, "naturally" suitable for crop cultivation, livestock herding, and settlement because of, e.g., poor agricultural soils, dense vegetation, harsh climatic conditions, and pests, plagues, and predators. Ovamboland is not unique in this respect. For example, a substantial part of the African landmass consists of deserts with extremely low rainfall. Even the environments that are better endowed with rainfall and water (Africa's forests, e.g.) require substantial investments of labor, capital, and other resources before they become suitable for agriculture. Temporary and partial clearing with fire (e.g., through shifting cultivation) was practiced across Africa as well as in Asia and the

was no incentive for its population to really alter the environment. See Scarborough, *The Flow of Power*, p. 15.

58 See, e.g., Scott, *The Art of Not Being Governed*, pp. 64–97.

59 W. Rodney, *How Europe Underdeveloped Africa* (London: Bogle-L'Ouverture, 1983 [1972]) and J. Miller, *Way of death: Merchant Capitalism and the Angolan Slave Trade, 1730–1830* (Madison: University of Wisconsin Press, 1996). See also Scott, *The Art of Not Being Governed*, pp. 64–97.

Americas. Shifting cultivators usually returned to and highly valued previously cleared forest lands after a period of fallow because such lands were easier to clear. In fact, previously cleared land, even if it was covered by secondary forest, was often considered the preserve of the clearing household or linage. The example of forest clearing in central-south Ghana in the sixteenth and seventeenth centuries dramatically highlights the significant investment that forest clearing required: it was accomplished through the *import* of African slaves, who were purchased with gold from European traders. The resulting environmental infrastructure facilitated the development of the powerful Akan city-states. The clearings became a major source of contention between increasingly powerful kingdoms, with Asante gaining domination over the Gold Coast and its hinterland in the eighteenth and nineteenth centuries.[60] In the same vein, social and political struggles in the Congo River Basin in the eighteenth and nineteenth centuries and the pre-1500 Great Lakes region, e.g., were over the control of "useful land," including cleared land with farms and villages, banana groves, gardens, and ponds (i.e., environmental infrastructure rather than about land or people per se).[61]

As the Ovamboland case illustrates, in semiarid environments, settlement, and agriculture required not only clearing vegetation but also constructing and maintaining an extensive water management and water harvesting infrastructure, ranging from large-scale to small-scale irrigation, to raised fields and drainage systems, and water reservoirs and wells. Examples of societies that developed similarly extensive environmental infrastructure include the Maya and other societies of Central America and the farmers of Java and Bali in Southeast Asia.[62]

Land with environmental infrastructure, including clearings for settlement and grazing, dry-season water supplies, fields for wet or dry agriculture, living quarters and food stores, orchards, gardens, groves,

60 R. Kea, *Settlements, Trade, and Polities in the Seventeenth-Century Gold Coast* (Baltimore, MD: The Johns Hopkins University Press, 1982). On slave plantations, see, e.g., S. W. Mintz, *Sweetness and Power: the Place of Sugar in Modern History* (New York: Viking, 1985).

61 See Schoenbrun, *A Green Place*, pp. 72–82, 92–93, 127–216. Harms makes a similar argument, see R. Harms, *Games against Nature: An Eco-Cultural History of the Nunu of Equatorial Africa* (Cambridge: Cambridge University Press, 1987).

62 See Whitmore and Turner, *Cultivated Landscapes of Middle America on the Eve of Conquest*, pp. 165–227, and Christie, "Water and Rice in Early Java and Bali," pp. 235–258. Annual rainfall on Java and Bali is much higher than that in Ovamboland, but Java, Bali, and much of Southeast Asia (and the same is true for much of Central America) are subject to distinct dry/wet season cycles that cause seasonal water deficiencies.

and coppice stands were so valuable that the inhabitants of Ovamboland (like societies elsewhere) made major investments in defensive structures. Fences and palisades, raised beds, mounds and walls, and dikes, ditches, moats, and canals of varying dimensions provided protection against human and animal predation, floods, and fire.

Environmental infrastructure was a source of wealth that could be used to attract and reward followers, who in return offered the labor and expertise to reproduce, maintain, and protect the resource. The allocation of partial (and temporary) rights meant that followers effectively became shareholders in the venture of environing, creating multiple rights to single resources, a phenomenon referred to as "bundles" of rights. For example, the rights to harvest fruit from trees on a plot could be awarded to X, the cultivation rights to the land to Y, and the main owner of a water hole or well on the same land could be Z, even though X, Y, and neighbors A, B, and C might share the use and maintenance of Z's water hole. As a result, resources sometimes were not only co-owned and (re)distributed along lines of kinship, but also through the creation of artificial kin relationships and patronage. Thus, ownership and access to environmental infrastructure resources was shaped by, and in turn shaped, social networks. Clan ownership of land and other resources was not entirely a colonial invention of tradition, but also developed from the circumstance that the production and reproduction of environmental infrastructure was a consummate social and socializing process. Here again, the example of Ovamboland in the twentieth century, where individuals and collectives (households, villages, clans) engaged in environing in highly differentiated ways was not unique. Environing in polders in China and the Low Countries in the tenth through eighteenth centuries (as in the irrigated Nile Valley between the sixteenth and eighteenth centuries) involved an array of individual and collective and state and non-state actors, often in intricate, highly dynamic combinations.[63] Multiple rights to land and resources are, of course, not anathema to "modern" society. Even in the haven of private property (the United States), property rights are

[63] In China, the history of land reclamation may date back as far as the twelfth century BC, but creating polders seems to have been principally a state project until about the tenth century AD. On China, see especially Mihelich, "Polders and the Politics of Land Reclamation in Southeast China during the Northern Sung Dynasty (960–1126)" and Stuermer, "Polder Construction and the Pattern of Land Ownership in the T'Ai-Hu Basin during the Southern Sung Dynasty." On the Low Countries, see TeBrake, Medieval *Frontier* and Verhulst, *Landschap en Landbouw in Middeleeuws Vlaanderen*. On the Nile Valley, see Mikhail, *Nature and Empire in Ottoman Egypt*.

not that absolute or exclusive. In fact, private land, e.g., is subject to a myriad of municipal, state, and federal restrictions and environmental and other regulations limiting how the owner can use (and abuse) his or her property and safeguarding the rights of others, including but not limited to rights of way and easements, conservation, pollution, public safety, and eminent domain.[64] Thus, both the Lockean idea that human labor investments create exclusive private property for the "owner" of the labor and the Marxian repost that labor creates communal property because labor cannot be alienated are problematic. Both are embedded in the Nature–Culture dichotomy and leave out important contributions to the processes of environing: they presume that "Nature" is inert and that it has no inherent value in and of itself. Moreover, both Locke and Marx focus solely on human labor investment, and not the contributions of, e.g., animal and plant inputs, or the impact of solar energy.

Conclusion

The inhabitants of north-central Namibia highly valued farms with dark-earth fields and fruit trees, water-harvesting holes and reservoirs, and coppice woodlands and savanna forages that had been shaped by browse and burning because the resources sustained their lives and livelihoods in a challenging and dynamic semiarid environment. The resources should be explicitly valued analytically and conceptually, as being neither entirely natural in origin and shape nor fully cultural. The inhabitants of north-central Namibia themselves did not consider the soils in their crop fields, their water holes, or their fruit trees to be natural givens. Evaluating and describing such resources as "natural" resources thus erase human agency and tens if not hundreds of thousands of years of Human–Nature interactions in Ovamboland, in Africa, and across the world. In contrast, valuing dark earths, fruit trees, water holes, coppice woodlands, and browse and burning regimes in the creation and maintenance of bushland savanna as environmental infrastructure acknowledges the extensive investments of people and other environmental actors in creating and recreating their environments.

64 On environmental and other regulations as affecting the rights of property owners, see Graham, *Lawscape*, pp. 168–173.

7

Science and the Failure to Conquer Nature

Environing and the Modern West

The previous chapters demonstrated how the environing concept recasts the analysis and description of environmental change in non-Western and premodern contexts beyond the Nature–Culture dichotomy by highlighting process over outcome and plurality over linearity. Non-Western and premodern societies neither lived by and in Nature nor had they subjugated Nature. Instead, human populations were engaged – with other environmental agents and forces – in continuous and multiple processes of environing. But the application of environing is not confined to the "non-West" and the "premodern." As a concept, it bridges the West–non-West and premodern-modern divides as well as the Nature–Culture divide. This chapter uses the post–World War II colonial project to conquer north-central Namibia's Nature to illustrate why and how the environing concept may be equally relevant to understand environmental change in the modern West.

After World War II, imperial governments initiated economic development projects throughout their African and Asian colonies, an undertaking that has been likened in intensity and impact to a second colonial conquest. In effect, rather than conquering territory and states, the main objective was to conquer non–Western Nature.[1] Colonial officials and experts deployed Western science and technology to dam rivers, drain

[1] The term "second colonial conquest" is Lonsdale's, see J. Lonsdale, "East Africa," in J. M. Brown and W. R. Louis (eds.), *The Oxford History of the British Empire, vol. 4: The Twentieth Century* (Oxford: Oxford University Press, 1999), pp. 530–544. For the second colonial conquest as a conquest of (non-Western) Nature, see Kreike, *Deforestation and Reforestation in Namibia*, pp. 47–81.

swamps, irrigate drylands, and overcome Nature's plagues: floods and famine, drought and disease, erosion, and desertification. Overall, the colonial project to conquer Nature was less than successful, even as it sometimes had dramatic impacts on colonial subjects and how they perceived and used their environments.[2]

The failure of colonial or imperial Western science and technology in the non-West is typically contrasted to its success in the conquest of Nature in the West, and it is ascribed to non-Western resistance or to the inherent incompatibility of Western science and technology with non-Western societies and environments. At issue is that the explanatory models derive from and reproduce the West–non-West and Culture–Nature dichotomies. Scott suggests an alternative model that moves beyond both the West–non-West and Nature–Culture binaries: he identifies imperialist state-makers and state-evaders as the engines of history. Imperialist state-makers sought to dominate Nature to make the environment "legible" to the state; that is, Nature was simplified to facilitate state extraction of taxes, labor, and other resources. In turn, state-evaders, used and recreated Nature to make their societies and the environments they relied on as "illegible" as possible. State-evaders enhanced the "friction" of their home territories by retreating into rugged mountains or swamps and by practicing hunting and gathering, nomadic pastoralism, and refuge agriculture, including shifting cultivation in a process Scott calls "self-barbarization."[3] Scott is pessimistic about the future of the few remaining stateless refuges that survive because their Natur(e)-alized environments are no longer an effective shield against the modern state's Cultur(e)-ization process of simplification.[4] But White's *Organic Machine* argument and the growing early twenty-first-century awareness that "Nature" neither has been fully understood nor entirely subjugated, suggests that the ability of the modern state-society to overcome Nature's friction has its limits in *both* the non-West and the West. One reason is that Nature appears to be far more resilient in the face of Culture's onslaught. The resilience usually is imagined in terms of a cornered wild animal lashing out at its hunter or as a higher spirit (God, Mother Nature) or force (Nature) unleashing punishing plagues on erring humanity. But the perception of Nature's resurrection and revenge as evidenced in floods, plagues, and global climatic doom is perhaps primarily the product of

[2] A recent example is Biggs, *Quagmire*.

[3] See Scott, *Seeing Like a State* and Scott, *The Art of Not Being Governed*.

[4] Scott, *The Art of Not Being Governed*, pp. ix–xi and 324–337.

having prematurely proclaimed Nature's death and the final victory of Culture. In fact, Nature has proven to be far more dynamic and, therefore, far more difficult to "master" than we imagined in the past century. Because "Nature" is so dynamic, using and managing "Nature" is never an outcome, but always a process that can be better appreciated in terms of environing. High modernism perceived Nature as passive and the domestication of Nature as a one-time act that once set in motion – either in the motherland or the colonies – would inevitably and irreversibly result in the conquest of Nature. But the premise proved erroneous, first in the colonies, next in the Western heartland, and increasingly so at a global level. Thus, Western colonial efforts to conquer non-Western Nature in the twentieth century may be more relevant than has been assumed to understanding the challenges that twenty-first-century modern societies face with Nature resurgent.

The South African colonial administration of Namibia participated in the larger imperial project of development through the conquest of Nature. In fact, by the late 1970s, colonial projects had substantially reshaped Ovamboland's environment. A hydraulic system with pumps, canals, a purification plant, and pipelines that supplied water from the Kunene River throughout the year made urbanization possible, provided clean water through a pipeline system, and supported small-scale irrigated horticulture and tree nurseries. The tree nurseries supplied seedlings to large eucalyptus plantations intended to combat deforestation by making available an alternative supply of construction materials and fuel wood for Ovamboland's growing population.[5]

Colonial officials and experts, inspired by modernization zeal, regarded the hydraulic system and the forest plantations as scientific tools to liberate Ovamboland from the constraints of a semiarid environment: water and wood would be in abundant supply despite deforestation and desertification. The eucalyptus plantations, however, contributed little to satisfy local wood needs. Moreover, a severe drought during the early 1990s demonstrated that scientific hydraulic technology had failed to make Ovamboland immune to the vagrancies of a semiarid climate. The large majority of Ovamboland's population had no access to the new water system, and some of those who were connected to the standpipes found them empty. Western science and technology therefore had not brought about the domination of Culture over Nature. But, colonial officials and

[5] On colonial water infrastructure projects elsewhere in Africa, see Tempelhoff (ed.), *African Water Histories*.

experts did contribute to creating environmental infrastructure. The history of colonial attempts to subjugate Nature in north-central Namibia suggest that the concept of environing applies to modern Western contexts as well as to premodern and non-Western contexts and highlights once again that environmental change should be understood as plural processes that each may be subject to their own chronologies even as they interact. Colonial water experts intentionally located dams and wells on the sites of preexisting indigenous water harvesting systems, thus grafting colonial infrastructure onto ongoing processes of environing, at the same time that they initiated new processes. Again, these processes occurred in many other places. For example, in the Kuruman region of South Africa, small-scale irrigated fields established by local farmers were appropriated by the colonial authorities and expanded and improved. Throughout South Asia, colonial administrations alienated and expanded on indigenous water infrastructures and reclaimed neglected or abandoned pre-colonial hydraulic works.[6]

Colonial Science and Environmental Planning

During the early 1940s, the Native Commissioner of Ovamboland concluded that the Ovambo floodplain area was overpopulated. He consequently postulated that serious environmental degradation, including deforestation, could be prevented only if the region east of the floodplain could accommodate further population increase.[7] The enormous stretches of land in eastern Ovamboland and south and southwest of the floodplain, however, could not effectively be settled because of a lack of water. In fact, a steady supply of water could not be maintained even for the small administrative staff at the Ovamboland colonial headquarters at Ondangwa in the floodplain, despite the use of modern dam and borehole technology. In 1941, the drinking water for the colonial staff at Ondangwa had to be rationed because the reservoir was empty "and water for laundering and for watering fruit trees etc. has to be fetched from a native owned water-hole, which was recently opened with the assistance of the

[6] Jacobs, *Environment, Power, and Injustice*, pp. 151–169; M. M. Islam, *Irrigation, Agriculture, and the Raj Punjab, 1887–1947* (New Delhi, India: Manohar, 1997); A. Agarwal and S. Narain (eds.), *Dying Wisdom*; D'Souza, *Drowned and Dammed*, pp. 38–41, 108–115. In the Mekong delta of Vietnam, French engineers also used preexisting hydraulic infrastructure, see Biggs, *Quagmire*.

[7] NAN, A450, 10, draft annual report Ovamboland for 1942 and a page marked with number: 56.

Government Native Staff." The situation at the Oshikango office was hardly better: its staff relied on a well, but its water supply diminished and turned brackish during the dry season.[8] In 1947–1948, plans to move the administrative headquarters at Ondangwa to another location with a more secure water supply were discussed, but the idea came to naught. In early 1952, after poor rains, Native Commissioner Eedes feared that his headquarter's water supply would run out toward the middle of the year: "[a]rrangements will have to be made to transport drinking water from Oshigambo or evacuate the European women and children."[9]

Ironically, and underscoring how the colonial project was at the mercy of the "forces of Nature," a rainy season with heavy local rains or heavy rains upstream in the watershed (which fed the semiannual flood or *efundja*) often cut off the Ondangwa headquarters from the rest of Ovamboland and repeatedly severed all overland communications with the rest of colonial Namibia. In April 1937, local rainfall and floodwater from Angola was so heavy that the Ekuma seasonal water course could no longer be traversed by vehicles.[10] In January 1944, heavy rainfall cut off Ovamboland from the remainder of the colony to all but radio traffic for three months and made travel to Angola impossible. The regular bus service from Ondangwa to Tsumeb to the southeast of Ovamboland only resumed in May and the children of the Church of England Mission missionaries who had visited their parents at St. Mary's at Odibo (close to Oshikango) for the holidays could not return to their boarding schools in Windhoek in time.[11] When Eedes succeeded Hahn as the Native Commissioner for Ovamboland in early 1947, one of his priorities was the building of all-season roads because "[w]ithout these roads Ovamboland cannot progress, or be effectively controlled, and no constructive works can be undertaken."[12]

The 1950 flood served as yet another harsh reminder of how vulnerable the South African colonial order was to the whims of Nature. During

[8] NAN, NAO 38, NCO to CNC, Ondangwa, August 14, 1941, and NAO 23, ANC to NCO, Oshikango, September 28, 1941.

[9] NAN, NAO 101, [Secretary] SWA to Administrator SWA, [Windhoek], February 2, 1948, and NCO to Director Works, Windhoek, November 22, 1947; NAO 60, Quarterly Report Ovamboland, January–March 1952.

[10] NAN, NAO 20, Monthly Report Ovamboland, April 1937.

[11] NAN, NAO 11, Annual Report Finnish Mission Ovamboland 1944 and NAO 21, Quarterly Report Ovamboland, January–March 1944; MacDonald Diary, Typed Mss. Period: 1932–1944 (18 pp.) (Private Collection Nancy MacDonald), courtesy Nancy MacDonald, Odibo, March 8, 1993.

[12] NAN, NAO 60, Quarterly Report Ovamboland, January–March 1947.

the 1949/1950 rainy season, the watercourses were so full that communications between the colonial offices at Ondangwa and Oshikango were cut off for months, long after the rains had ceased. The report for the first quarter of 1950 paints the image of an imminent reconquest of Ovamboland *by Nature* with water eroding the very foundations of the colonial order, figuratively speaking as well as literally (as it undermined buildings and roads and destroyed government food stores and other colonial infrastructure):

> the whole of Ovamboland . . . [was] inundated in March. The standing crops are in great danger. . . . The following is another note received from a Messenger in Ongandjera: 'The corn baskets have fallen down. Corn is rotten and wasted. Fishes and sardines can be seen in the [colonial government] camp and in the [colonial] corn depot'. . . . Crocodiles are reported to have appeared in the Onkolonkathi area in the North West.[13]

In 1951, most roads within Ovamboland and from and to the district were again cut off during part of the rainy season. Another major flood in 1954 isolated Ovamboland for almost four months and severely hampered labor recruitment efforts for the Namibian and South African farms, mines, and industries.[14]

The administration marshaled all its scientific resources to conquer Nature in Ovamboland and prevent a repeat of the situation in 1950. The Department of Works and the Native Commissioner for Ovamboland proposed a scheme of combined road and water conservation projects to address water shortages during the dry season and flooding during the rainy season. Engineers initiated a pilot project at Okapya, nine miles north of Ondangwa, where the road crossed one of the major seasonal watercourses between Ondangwa and Oshikango. The project called for a dam to be constructed using soil excavated from the upstream side of the watercourse. The construction would permit water to be stored up to one meter (three feet) above the bed of the watercourse. The dam also functioned as a traffic lane above the rainy season flood level.[15]

[13] NAN, WAT 144, Secretary [Department of Works?] to Administrator SWA, [Windhoek], December 4, 1950 [*sic*? 1951?] and Director of Works to CNC, Windhoek, August 30, 1950. See also NAO 60, Telegram NCO to Sec. SWA, Ondangwa, February 17, 1950, and Quarterly Report Ovamboland, January–March 1950.

[14] NAN, NAO 60, Quarterly Reports Ovamboland, January–June 1951 and NAO 65, Annual Health Report Ovamboland 1954.

[15] NAN, WAT 144, Secretary to Administrator SWA, [Windhoek], December 4, 1950 [*sic*? 1951?] and Director of Works to CNC, Windhoek, August 30, 1950.

Next to the water storage dams, boreholes were the main technology used to attempt to harness Ovamboland's "natural" and "wild" water resources through the 1960s. A 1947 report on the geology and groundwater conditions concluded that boreholes that relied on deep fresh water aquifers could bring relief to the eastern and western parts of Ovamboland. In the central areas, which constituted the heart of the densely settled floodplain, however, experts deemed boreholes to be inefficient because of highly saline aquifers. Instead, the Director of Works proposed an improved version of the excavation dams that had been introduced through food for work projects during the 1929–1931 Famine of the Dams. Significantly, the Director of Works, an engineer, concluded that excavation dams were the only useful addition colonial science had to offer to the indigenous water hole technology. The excavation dams, however, had two drawbacks. First, the dams could not be excavated too deeply for fear of contamination from subsurface saline water layers. Second, the dams were small, and they were rapidly polluted by people and animals. As an alternative, the Director of Works proposed excavation storage systems, which consisted of star-shaped series of trenches excavated in such a way that they sloped slightly toward the center of the star. A concrete well with sand strainers was placed in the center and the trenches were filled with loose sand. Part of the water would be soaked up by the sand, which served to reduce evaporation and to filter the water.[16]

Between 1947 and 1953, however, little came of these plans. Engineers successfully drilled four boreholes in eastern Oukwanyama beyond the floodplain and equipped them with pumps and engines. They also drilled a number of boreholes at and near Oshigambo mission, but the water was of questionable quality. Two boreholes constructed in Uukwaluthi district in western Ovamboland yielded water unfit for human consumption. The construction of a dam at the Ondangwa administrative offices was a failure. Only the pilot dam causeway north of Ondangwa on the Ondangwa-Oshikango road functioned, and, although the floodwaters destroyed the causeway early in 1953, thirteen additional dam-causeways were envisaged between Ondangwa and Oshakati, and sixty-six from Ondangwa westward to Uukwaluthi and Tshandi. Pipes through the dam-causeways

[16] NAN, NAO 101, Director of Works, "Ovamboland: Geology and Groundwater Conditions," November 5, 1947. McKittrick somewhat overestimates the depth of the original dams when she writes: "today it is difficult to recognize in these shallow depressions the deep squared basins they once were." See McKittrick, "The Wealth of These Nations," pp. 449–469 (especially 463).

allowed surplus water to flow to the downstream side of the dam. Colonial experts proposed that water erosion could be reduced by temporarily protecting the sides of the dam-causeways with a frame of small poles and sticks until a thick grass cover had developed.[17]

Colonial Artifice and Environmental Infrastructure

Colonial water reservoirs, known as "dams" in Namibia and South Africa, may not have substantially augmented the total volume of available water. Despite colonial claims to the contrary, the reservoirs typically merely replaced existing "indigenous" water hole complexes (singular *etambi*, plural *omatambi*). While the dams may have "harvested" more water, water preservation and distribution were far less efficient than in the case of the "traditional" water holes, and water losses through evaporation were considerably higher.

Between 1953 and 1966, the colonial administration constructed more than 300 water storage dams in Ovamboland. The dams typically had a capacity of 20,000–30,000 m³ (5–7.5 million gallons) and were often constructed at the location of preexisting indigenous water hole complexes or wells. In the early 1970s, e.g., an official requested the Tribal Council of Uukwaluthi district to identify sites with "water hole complexes" to pinpoint the best locations for the construction of colonial dams. A colonial dam replaced a water hole complex in at least one location in Uukwaluthi in that period. A small number of the colonial dams were pump storage dams: the stored water fed wells located on the side of the reservoir. The colonial administration only constructed pump storage dams at selected sites including its administrative headquarters at Ondangwa and Oshakati, near hospitals, e.g. at Odibo, and at "tribal" capitals. The well shafts in the pump storage dams consisted of brick and a concrete slab covered the top to limit pollution. Water use was expected "to be small and only for the utmost priorities; and in their own interest the population has to assist in ensuring that water is not wasted."[18] Food

17 NAN, NAO 101, NCO to CNC, Ondangwa, February 14, 1953, and WAT 144, "Report on the Details of Schemes," January 29, 1953.

18 NAN, WAT 148, Director Water Affairs to Bantu Commissioner, Windhoek, March 23 and 24, 1966, and BOS, District Record Book Oshikango [1965], pp. 10–11; OVA 33, Senior Magistrate Ondangwa to Superintendent Onandjokwe Lutheran Hospital, Ondangwa, March 4, 1970. For the selection of water hole complexes as the sites of colonial dams, see OVA 33, Agricultural Officer to Sec. Agriculture, Ondangwa, November 22, 1974 [or 1977?] and Agricultural Officer to Sec. Agriculture, Ondangwa

for work programs during the 1953 and 1958–1961 droughts provided cheap labor for constructing dams and other water projects.[19] From 1961 onward, however, dam construction was less dependent on local labor. A bulldozer or tractor purchased with Tribal Trust Funds was a new important tool for constructing dams. The bulldozer prepared the ground for all five dams constructed in 1962, as well as for one of the three dams completed in 1964, four of the thirteen dams that were added in 1965, and the single dam that was constructed in 1966. Private contractors constructed most of the other dams.[20] Some of the dams were intended to support the creation of planned towns.[21]

The new dams, however, proved less effective than had been anticipated. During the countrywide 1958–1959 drought, rainfall in most of Ovamboland was far below average and the flood (*efundja*) did not reach Ovamboland. A survey of eighty dams revealed that only 30 percent of the dams (mostly located in the eastern Ovamboland, where rainfall had been better) were full. Twenty-two percent of the dams were empty. All of the dams that were less than three-quarters full (i.e., 40 percent of the dams) were expected to run out of water within two to three months. Less than half of the dams, therefore, were expected to contain water during the height of the dry season. Generally, the rainy season did not bring relief until October or November, at the earliest. In western Ovamboland, people had deepened their "wells" (probably mostly water holes) as early as May in an attempt to find more water, an activity that typically took place toward the very end of the dry season.[22] Conditions were better in subsequent years but were far from satisfactory. Surface water supplies following the rains in late 1959 quickly ran out by early 1960. The large dam at Okatana contained a mere two feet (0.6 meters) of water, and most storage dams (except those in eastern Ovamboland) only contained

[November 22, 1974]. On the typical size of the dams, see WAT 148, Director Water Affairs to Chief Engineer Conservation Rhodesia, Windhoek, July 16, 1965. On pump storage dams, see WAT 147, Register Dams in Bantu Areas, Appendix no. 7 to Five Year Plan Water Affairs [1962 or 1963].

19 NAN, BAC 122, Chief Bantu Commissioner to Minister Bantu Administration and Development, Windhoek, June 26, 1959, and May 16, 1959.

20 NAN, BOS, District Record Book Oshikango, pp. 10–11. See also WAT 148, Director Water Affairs to Chief Engineer Conservation Rhodesia, Windhoek, July 16, 1965; Progress Report Dams, Bridges, February 28, 1966, and Director Water Affairs to Bantu Commissioner Ondangwa, Windhoek, March 24, 1966.

21 NAN, WAT 148, Adjunct Director Water Affairs to Director Works, [Windhoek], June 11, 1965 [?] and Sec. Bantu Administration to Chief Bantu Commissioner Windhoek, Pretoria, February 2, 1965.

22 NAN, BAC 122, Director Water Affairs Department to CNC, Windhoek, May 13, 1959.

muddy water. During this time, people reportedly relied only on "traditional" water holes. The staff of the South West Africa Native Labour Association (SWANLA) shop at Ombalantu hauled water all the way from the Kunene River and the Ondangwa government station.[23] In the 1961–1962 dry season, one of every five dams was completely dry.[24]

By 1968, Ovamboland had 257 storage dams and 45 dams of other types. Evaporation losses in the shallow excavation dams were high. Observations at Mahanene Research Center in Ovamboland in the 1970s yielded evaporation rates of nine to eleven feet (2.9–3.2 meters) per annum, with more than 30 percent of the annual losses occurring during the months of October, November, and December. Most dams were less deep than the annual evaporation was high. By the mid-1960s, the use of floating slabs in the pump storage dams may have cut evaporation losses by half. The slabs consisted of a mixture of cement, sand, and plastic. In the late 1960s, evaporation losses at the Oshakati dams were eight inches (20 cm), probably on an annual basis, but at the hospital dam, which was likely a pump storage dam with floating slabs, evaporation was cut in half to four inches (10 cm).[25] Another problem was the silt buildup (up to one to two feet each year), especially in the pump storage dams, although the clay buildup also may have sealed the reservoirs off from saline water infiltration.[26]

Colonial Water Technology and Environmental Change

By the early 1960s, borehole exploration had only proved successful in the southwestern and northeastern margins of Ovamboland; elsewhere, the water was too saline. In 1968, Ovamboland had a total of 65 working boreholes. Boreholes were costly, particularly given the high failure rate for discovering potable water. In 1968 alone, forty-nine drilling attempts had failed entirely. Of the fifty-four boreholes that struck water, only eighteen that yielded water potable for humans were made fully operational. The windmill-powered pumps required high levels of maintenance

23 NAN, BAC 123, Chief Bantu Commissioner SWA to Sec. Bantu Affairs, Windhoek, February 8, 1960.

24 NAN, WAT 146, SWA Administration, Windhoek, July 3, 1962. The table presents data on all the dams and causeway dams.

25 NAN, WAT 148, Director Water Affairs to Chief Engineer Conservation Rhodesia, Windhoek, July 16, 1965, and OVA 48, Venn, Loxton, Hunting & Associates, Periodic Progress Report Mahanene Research Station, June 15–16, 1974; WAT 333, Director Water Affairs to National Institute Transport, November 30, 1978. On the evaporation data, see WWA [WAT?] 640, Water Use October 12, 1967–February 1969.

26 NAN, WWA [WAT?] 636, Damlogs 1970 and Report J. D. Peterssen, July 23, 1969.

and the lack of spare parts at remote sites led to frequent breakdowns. Borehole yields varied from as little as 9 gallons (36 liters) per hour to more than 4,000 gallons (16,000 liters) per hour.[27] In 1974, a government geologist recommended halting borehole prospecting in Ongandjera in the southwestern side of the floodplain because the high salinity and fluoride levels made the water unsuitable for human consumption.[28] Boreholes continued to be constructed, however, and by 1980, Ovamboland contained a total of 206 boreholes, some of which were privately owned by chiefs and headmen.[29]

In the early 1960s, the colonial administration introduced a grand scheme to construct canals across the floodplain. The canals were to be oriented perpendicular to the north-south drainage of the floodplain and they initially were conceived to redirect the flood and rain waters to the envisaged administrative and core urban areas in Ovamboland. In the 1970s, the canals were recycled to redistribute water pumped in from the Kunene River. The plan called for the main canal to cover a distance of seventy to eighty miles (100 km) from the Kunene River to close to Okatana, about twenty-five miles (40 km) to the northwest of Ondangwa. A second canal served to guide water into the Etaka seasonal river that formed the southwestern edge of the floodplain. The administration initiated the project in 1959 and by early 1960, five miles (eight kilometers) of canal had been constructed near the large Okatana dam.[30] Water losses in the canal were expected to be as high as 21 percent of the total volume carried, partly through percolation, but mainly through evaporation. By the end of 1965, the canal ran from east of Ombalantu to the Oshakati Dam, which was then connected to Ondangwa through a pipeline. The

[27] NAN, WWA [WAT?] 222, Bantu Commissioner to Chief Bantu Commissioner Windhoek, Oshikango, August 15, 1967, and Chief Bantu Commissioner to Director Water Affairs, Windhoek, November 7, 1966; WWA [WAT?] 223, Report Boreholes Western Ovamboland, September 24, 1970; WAT 66, Sec. SWA to Sec. Bantu Administration Pretoria, Windhoek, June 21, 1961; AHE (BAC) 1/352, Annual Report Agriculture Ovamboland 1968; OVA 33, Adjunct Director Geological Survey to Director Water Affairs Windhoek, Windhoek, January 29, 1974.

[28] NAN, WWA [WAT?] 223, Report Boreholes Western Ovamboland, September 24, 1970; WAT 66, Sec. SWA to Sec. Bantu Administration Pretoria, Windhoek, June 21, 1961; AHE (BAC) 1/352, Annual Report Agriculture Ovamboland 1968; OVA 33, Adjunct Director Geological Survey to Director Water Affairs Windhoek, Windhoek, January 29, 1974.

[29] NAN, OVA 49, Sec. Agriculture to Sec. Department Chief Minister, Ondangwa, May 3, 1980; OVA 33, Ongandjera Tribal Authority to Sec. Agriculture, Okahao, April 4, 1978, and Sec. Agriculture to Magistrate Ondangwa, Ondanwa, May 21, 1979.

[30] NAN, BAC 122, Chief Bantu Commissioner to Minister Bantu Administration Windhoek, June 26, 1959, and May 16, 1959; BAC 123, Chief Bantu Commissioner SWA to Sec. Bantu Affairs, Windhoek, February 8, 1960.

Finnish Mission station and its hospital at Elim were connected to the main canal with a smaller canal. The Etaka canal was completed between Eunda and Ongandjera using part of the Etaka seasonal riverbed.[31] The canal required periodic cleaning to remove silt and plant growth.[32]

It is difficult to measure unambiguously the impact of changes in Ovamboland's water infrastructure brought about by colonial projects in terms of environmental improvement or degradation. But overall, environmental change was dramatic. First, as a whole, the pattern of the availability of water resources changed both seasonally and structurally, but it did so unevenly because the colonial hydraulic infrastructure had a limited reach. The canals permitted for greater densities of human and animal populations, even in areas where no human settlement previously had been possible. Because a pipeline ultimately connected the canal to the Kunene River, the canal subsequently held water throughout the year, even during severe droughts. In many areas, the canals, boreholes, and dams thus greatly improved water availability. In other areas, however, the effect was exactly the opposite: the new colonial water infrastructure decreased water availability. The canals – especially the Ovamboland canal from the Kunene to Okatana – cut diagonally through the southern floodplain drainage lines and quite effectively redirected floodwater from the seasonal watercourses southeast toward Okatana (and later Oshakati and Ondangwa).[33] The possibility that directing the water in the floodplain environment away from the areas south of the canal (including the Etosha Park) might decrease water availability was considered, but it was thought that colonial technology could easily address the situation. Moreover, the impact caused by the strategy has not been taken into account in later environmental analyses of Ovamboland and the Etosha Park. The issue was raised somewhat hesitantly during a 1961 "tribal" meeting in Uukwambi district; after the secretary of the meeting already had noted in the minutes "[n]o more speakers," a Paulus Siyemba warned: "[i]t will

31 NAN, WAT 159, H.J. Van Eck to Dr. O. Wipplinger, Johannesburg, February 21, 1963, and document following entitled "Estimated Costs of 1,800 Cusec Canal"; WAT 3, Sketch; WAT 147, Director Water Affairs Ovamboland, Canal Scheme Construction Program, 1963/64.

32 NAN, WWA [WAT?] 636, Senior Inspector N. H. Schwartz, Owambo Report on Waterinstallation, May 22–June 2, 1972.

33 NAN, WAT 147, Director Water Affairs Ovamboland, Canal Scheme Construction Program 1963/64. See also WWA [WAT?] 640, Director Water Affairs, Memo Watersupply Capacity of Oshakati-Ondangwa Scheme, July 15, 1966; WWA [WAT?] 637, Owambo Master Plan: Notes of Meeting Water Affairs with Commissioner-General, Oshakati, March 15, 1972.

be difficult because the canal has cut off the water."[34] The Bantu Affairs Commissioner for Ovamboland took note of this warning and reported to his superior in Windhoek that "the inhabitants of those areas where canals have been dug to lead water to storage dams... have objected because their water in the pans and storage/gathering places nearby their huts is led away." The Director of Water Affairs for colonial Namibia conceded that the canal, which redirected the floodwaters descending from the north, effectively functioned as a dam that prevented water drainage further south, and he recommended that sluices be constructed at all the eighteen flood channels that the canal cut through to allow water to flow south of the canal when required.[35]

Although work crews installed pipes through the causeway dams to allow floodwater to pass, the dams at best delayed the passage of water downstream. The roads themselves were intended to ensure all-weather communications and had been built above the high-water mark to facilitate water storage upstream and in practice functioned as a series of dams in the flood channels. Between 1953 and 1958, workers constructed sluices at approximately fifty locations along the road to allow the water to pass through. The road, however, continued to be damaged by floods. Between 1964 and 1967, the administration upgraded the Oshakati-Ombalantu road to a gravel road with concrete bridges and fords to improve all-weather access to the canal as well as to protect the canal itself against flood damage. The effective damming of at least eighteen floodwater channels by the road/canal combination, however, also resulted in higher losses through evaporation and percolation upstream. As a result, the amount of floodwater downstream of the Ovamboland canal and the Ondangwa–Ombalantu–Ruacana road decreased.[36] On

34 NAN, BAC 44, Minutes of the annual and quarterly meetings held at Uukwambi on June 15, 1961.

35 NAN, WAT 146, Bantu Affairs Commissioner to Chief Bantu Commissioner Windhoek, n.d., Appendix to Chief Bantu Commissioner SWA to Director Water Affairs Windhoek, Windhoek, July 11, 1961; WAT 147, Director Water Affairs Ovamboland, Canal Scheme Construction Program 1963/64. A similar objection was raised in the 1930s against the construction of the Jonglei Canal in the Sudd Marshes of the Sudan. The fear was that the canal would block east-west drainage patterns and divert water away from other areas in the south. And, once it had been partially dug, the canal, which had a dike on one side with an all-weather road on top, caused flooding on one side and a decrease in water availability northwest of the canal. See Collins, *The Waters of the Nile*, pp. 187, 365–367.

36 W. Bertelsmann, "Wasserbau im Ovamboland," *SWA Annual*, 1959, pp. 141–144; WWA [WAT?] 456, Acting Chief Road Engineers to Administrator-in-Executive-Committee, November 22, 1967.

the other hand, because downstream run-off was obstructed, flood levels may have increased directly north of the canal, potentially damaging crop fields. Moreover, silt deposited upstream of the canal/dam may have clogged up the inlets to pipes and sluices and may also have caused fields to flood.[37]

Drought conditions during the late 1960s proved once again that the colonial water infrastructure had failed to ensure the permanent domination of Culture over Nature: the dams and the canals were unable to sustain the increased demand for water, especially in the new towns that emerged along the main canal. In 1967, trucks hauled water to the Oshakati area, where taps had been closed because of the lack of water, and water was sold to the inhabitants of the rural areas around Oshakati in drums. Illegal water tapping may have accounted for more than one-third of total "consumption"; a 1970 report noted that the main gauge indicated a total water consumption of more than 15,000 cubic meters yet the total registered by user gauges amounted to less than 10,000 cubic meters. By the end of 1970, the Oshakati-Ondangwa "complex" was so short of water that it had to be rationed. The emergency pipeline that the administration constructed along the road from Angola to Oshakati initially was originally intended to be a temporary solution.[38]

The colonial water system was expanded and upgraded in the late 1970s. Construction was begun on a permanent pipeline between Ogongo–Oshakati–Ondangwa in 1976, and the pipeline feeding Kunene water from the Ruacana dam to the canal was upgraded in 1977. At the same time, the administration added a fish-grate pipeline system that was extended to areas close to the Ondangwa–Oshikango road, including Omafo, Eenhana, and Omungwalume, to facilitate relocating war refugees from Angola from camps along the border.[39] As with the canal, the availability of a permanent water supply permitted a much higher density of people and animals than before. This factor is very clearly reflected in the settlement patterns that developed following the construction of the

[37] On siltification processes and flooding, see, e.g., B. Dalal-Clayton, "Environmental Aspects of the Bangladesh Flood Action Plan." International Institute for Environment and Development (IIED), Issues Series No. 1 (1990), p. 11.

[38] NAN, WWA 640 f. 31/3/2/1, Report by H.W. Stengel, October 23, 1967; OVA 33, Director Water Affairs to Chief Director Owambo Government, Ondangwa, February 4, 1970; WWA 644, Director Water Affairs to Chief Roads Engineer, Windhoek, February 19, 1971.

[39] NAN, OVA 40, Telex Sec. Water Affairs to Director [Water Affairs], Windhoek, June 18, 1976; WWA 640, Report C. Bon, Ovamboland Pipelines, October 1977; WWA 637, Sec. Agriculture to Sec. Department Chief Minister, Ondangwa, December 2, 1976.

canal and the pipeline: these areas are presently the most densely settled locations in Ovamboland. They are also the areas where the use of other environmental resources (woody vegetation and grazing) is the heaviest, resulting in local shortages. Water demand surpassed supply during the early 1990s; in the dry seasons, the pressure in the pipeline system was so low that the stand pipes at its northern and southern extremities at Odibo and Okatana, respectively, ran dry.[40] In 1980, Ovamboland had 393 livestock dams, 130 reservoir dams, 106 miles (160 km) of pipeline, 206 boreholes, and 95 government wells.[41]

The history of water and water management in north-central Namibia, however, is neither a linear narrative about the victory of Culture over Nature nor a linear narrative about the decline of indigenous knowledge and practices in the face of Western science and technology. In fact, the 1991 census found that only 24 percent of Ovamboland's 100,000 households relied on piped water; a mere 9 percent drew water from boreholes; and 8 percent obtained water from the canals. The remainder relied on water holes and wells.[42] Moreover, "indigenous" knowledge proved to be far from static or merely *resistant* to innovation. Well technology, e.g., was rapidly integrated into Ovamboland's bodies of indigenous technology and knowledge. While water holes are an unambiguously "indigenous" technology, categorizing wells is more challenging. First, well technology and its knowledge base in general are neither a Western monopoly nor invention, and the technology is not a product of modern Science. Second, before the colonial era, wells were rare in Ovamboland. Third, well technology spread rapidly in Ovamboland in the colonial era after local well builders adopted it to local environmental conditions.[43]

Colonial Science and Plantation Forestry

Colonial science also presented a solution to the increasing pressure on Ovamboland's *natural* woody vegetation: plantation forestry. Experiments in the 1960s to identify suitable species led to the establishment of large eucalyptus plantations in Ovamboland during the 1970s and

[40] Personal observations, author, 1991–1993. A study in the western constituencies of eastern Ovamboland showed that population densities especially increased close to the pipeline system, see Erkkilä, "Living on the Land," p. 72, figure 40.

[41] NAN, OVA 49, Sec. Agriculture to Sec. Department Chief Minister, Ondangwa, May 13, 1980.

[42] Namibia, 1991 Census, Report A, Statistical Tables, vol. v, table H06.

[43] See Kreike, *Re-creating Eden*, pp. 143–154.

1980s. In contrast to the experience with water and water management, colonial foresters made little or no effort to build on or make use of any local knowledge regarding the propagation and management of trees and forests or carefully assess local needs. Instead, whereas colonial foresters remained fixated on the conventional forestry concerns with timber-quality eucalyptus trees produced from seeds, local needs were overwhelmingly for poles and sticks that were more easily produced through coppice. Ironically, the seed imported from South Africa, was from a very narrow genetic base and of poor quality.[44]

In 1964, Ovamboland was assigned its own forester. Initially, the forester primarily assisted in setting up the commercial exploitation of the indigenous Transvaal teak, a relatively small-scale undertaking that nevertheless, by the late 1960s, led to the cutting down of timber quality trees as far east as Eenhana, beyond the floodplain.[45] Small-scale tree planting trials to identify viable plantation species took place at different sites in Ovamboland throughout the 1960s. In 1960, the colonial administration established an irrigated perimeter and dryland plots at Okatana for regular crop trials. In May 1961, colonial staff planted eight treelings each of seven species, including *Eucalyptus camaldulensis*, *Prosopis juliflora*, and *Casuarina cunninghamiana*. In addition, the salt-tolerant fodder bush *Atriplex nummularia* was planted in seedbeds in August and transplanted in January/February (during the rainy season). By August 1961, only *Prospopis*, *Casuarina*, *E. camaldulensis*, and *Atriplex* did well enough to warrant their use in subsequent larger-scale trials.

In January 1962, workers planted 337 new trees (apparently treelings as opposed to seeds) at Okatana; eucalyptus (including *E. camaldulensis*) constituted almost two-thirds of the small-scale trial that involved a maximum of thirty treelings per species. Despite weekly watering and the application of dieldrin to combat insect infestation, few treelings survived. *E. camaldulensis* stood out favorably; more than half of the

44 See R. N. Cromer and K. G. Eldridge, "The Eucalypts as Tree Crops," F. T. Last (ed.), *Tree Crop Ecosystem* (Amsterdam: Elsevier, 2001), pp. 229–269.

45 NAN, AHE (BAC) 1/343, District Forest Officer, "Report on pitsawing operations in the Northern Bantu Territories," Grootfontein, April 1, 1965, and 1/352, Annual Report Agriculture Oukwanyama for 1964; AHE 1/342, Native Agriculture: Trust Forests: Main File, Quarterly Report District Forest Official of the SWA Administration, June 30, 1967, and Annual Reports District Forest Official of the SWA Administration for 1965/66 and 1968/1969 and Quarterly Report, June 31, 1966; BAC 132, Agriculture Officer to Bantu Commissioner, Grootfontein, July 30, 1961, Report on Work on Okatana Irrigation Scheme, July 30–September 11, 1961, and Quarterly Report Agriculture Ovamboland, June 30, 1958.

treelings of the species survived and the district forester enthusiastically recommended its introduction in Ovamboland.[46]

Subsequently, in March 1964, halfway through the rainy season, a further 370 treelings (including *E. camadulensis*) were planted at Okatana and at a new location at Oshikango. Drought, termites, and frost conspired to wipe out both the Okatana and Oshikango trial plantings in the next four years. Undeterred, the forester established a small irrigated experimental eucalyptus plantation at Oshikoto Lake just south of Ovamboland in 1966. Two years later, *E. camaldulensis* was one of four species planted at the new Oponono Lake experimental plantation south of Ondangwa. Again, however, the trial was very small-scale: only sixty-four treelings of each species were planted. The eucalyptus did reasonably well.[47]

The relative success of eucalyptus encouraged larger trials with the species to select provenances suitable for the environmental conditions in north-central Namibia that would assure a future supply of construction and firewood for Ovamboland. An additional objective of the trials was to test their potential as sources of nectar for bees, because experts considered Ovamboland to be short of natural food for bees.[48] The directorate of forestry in the 1970s established two large forestry plantations in Ovamboland: at Onuno and at Onankali. A third, much smaller forestry plantation at Leeukop along the Ondangwa-Tsumeb road in the far south of Ovamboland, was the location of a much smaller series of contemporary trials that were less scientifically rigorous and significant.

The first provenance trial established at the Onuno plantation with *E. camaldulensis* and *Eucalyptus tereticornis* started in 1973 and formed

46 NAN, BAC 132, sketch map "Section of Okatana where trees will be planted" appendix to District Forester to Chief Bantu Commissioner, Grootfontein, October 6, 1960; Agriculture Officer to Bantu Commissioner, Grootfontein, July 30, 1961, Report Okatana Irrigation Scheme, July 30–September 11, 1961; Chief Bantu Commissioner to Bantu Commissioner Grootfontein, Windhoek, August 3, 1961; Agriculture Officer Grootfontein to Bantu Commissioner, Grootfontein, July 31, 1962 and Chief Bantu Commissioner to Bantu Commissioner Ondangwa, [Windhoek], September 4, 1962; Agriculture Officer Grootfontein to Bantu Commissioner, Grootfontein, July 31 and September 25, 1962; AHE 1/342, Native Agriculture: Trust Forests: District Forest Official, Quarterly Report, December 31, 1963.

47 NAN, AHE 1/342, Native Agriculture: Trust Forests, District Forest Official SWA Administration, Annual Reports 1965/1966 and 1968–1969; Quarterly Reports, March 31 and December 31, 1964; June 31, 1966; and December 30, 1968.

48 E. Kreike, "An Inventory of Trials with Exotic Tree Species in Northern Namibia, with Special Reference to Provenance Trials with Eucalyptus spp" (Windhoek: Directorate of Forestry Internal Report, 1992).

part of a series of provenance trials with *eucalyptus* throughout Namibia and South Africa. The Onuno plantation was located along the Ondangwa-Oshikango road, just south of Oshikango on deep sandy soil.[49] The first treelings, grown from seeds planted in September 1973, were transplanted to the Onuno plantation in the middle of the rainy season of 1974. Although reports are contradictory, it seems that some soil preparation took place prior to planting, and that plastic covers and fertilizer were applied selectively. The plot was laid out in a conventional rectangular lattice pattern with three replications; each plot held five rows of five trees. The trial was abandoned in 1982.

A contemporary trial at Onuno experimented with *Eucalyptus citriodora* and eucalyptus hybrids but was very poorly executed from the start: the planting followed no particular design, no replications were used, and the plots were irregular.[50] As a result the trial's scientific value was negligible. A second series of trial plantings of *E. camaldulensis* and *E. tereticornis* was inaugurated during the 1977/1978 season with seed received from Pretoria. Twelve provenances of *E. camaldulensis* and three provenances of *E. tereticornis* were seeded in the nursery in August 1977. Not by design, but because of a shortage of tools and inputs, the soil was not prepared and no fertilizer was applied. The seedlings that survived were planted at Onuno during the rainy season in early March 1978.

A 1989 survey identified 238 surviving trees from the 1973/1974 trials at Onuno. The best eucalyptus treelings had grown to a height of approximately twenty-four feet (8 meters) and a diameter at breast height of approximately five inches (12 cm). Thus eucalyptus proved capable of growing to a size that could be used as poles to construct "traditional" huts, fences, and palisades in ten to fifteen years but the spindly trunks were a far cry from the timber quality output that foresters had imagined.

In 1976, staff established a far larger production plantation at Onankali, southeast of Ondangwa on the Ondangwa-Tsumeb road. This was probably the site of the short-lived colonial mopane experiments. On the basis of the experience with the Onuno trials, *E. camaldulensis* and *E. tereticornis* were the principal species selected for a series of plantings in the late 1970s on the western side of the road in what was

[49] Archives of the Forestry Officer Ondangwa at Ondangwa, file "forestry 1/5/3 Oununo woodlots (trial)," J. L. van Wyk to Secretary for Forestry Pretoria, Ondangwa June 8, 1977. The trial is known as Ovambo/1.

[50] The trial is known as Ovambo/2.

called Block A. In 1976 and 1977, more than 5,000 seedlings were planted at regular spacings. The growth rates proved to be very similar to those acquired at the Onuno trial plantation.

The 1977/1978 series at Onankali were much larger, involving almost 12,000 plants with somewhat wider spacing than used at Onuno and involving the same species. A smaller percentage of the treelings were grown from seed collected from the 1973/1974 provenance trial trees at Onuno; the remainder originated from seed imported from South Africa and supplied by Pretoria.

A further series planted the next year involved almost 15,000 seedlings, principally *E. camaldulensis* and *E. tereticornis*. In 1979, of the more than 30,000 treelings planted at Onankali's block A during the late 1970s, only 5,700 remained, suggesting a survival rate of 19 percent. After ten years, the surviving trees showed growth rates identical to those measured during the almost contemporary Onuno trials.

Later trials were planted on the portion of the Onankali plantation on the eastern side of the road. Almost 39,000 treelings were planted in 1979 and another 17,000 in 1980. The principal species in the 1979 planting – *E. camaldulensis* – did well; a 1989 survey demonstrated a survival rate of 80 percent, but the other eucalyptus species from the trial had been all but wiped out. Height growth rates measured for these series were somewhat better than in the earlier series but the diameter growth was comparable. The last plantings at Onankali took place in 1982 but were much smaller in scale, involving a mere 1,000 treelings.[51]

In his report about the 1973/1974 eucalyptus provenance trials in Namibia and South Africa, the agricultural official Kevin Darrow concluded that "[i]n the dry to arid regions . . . Eucalyptus camaldulensis was superior in survival and growth to Eucalyptus tereticornis, although neither showed satisfactory development." He emphasized that irrigation or accessible groundwater reserves were critical conditions for the success of eucalyptus plantations. He judged height growth in Ovamboland after forty-four months as very good (highlighting that the best lots exceeded twenty feet or six meters in height), especially compared to a mean height of only six feet (1.8 m.) in the Okavango region to the east of Ovamboland. But Darrow also concluded that stem diameter growth in what he called semiarid Zone 2 (which included northern Namibia) was poor, and although performance in the Ovambo trials was much better than in any

[51] For a more detailed description of tree trials in north central Namibia, see Kreike, "An Inventory of Trials with Exotic Tree Species in Northern Namibia."

other Zone 2 trial, even the Ovambo trials' results in this area attained only a fraction of the performance in the less arid Zone 1 regions involved in the trial. Moreover, stem forms of both subspecies in either zone were typically poor to bad: many stems were too crooked to serve as timber or telephone poles. For northern Namibia overall, where the ground water table was relatively high, Darrow recommended *E. tereticornis* because it outperformed *E. camaldulensis* in growth and stem form, the latter clearly reflecting the extent to which South African experts were preoccupied with growing timber quality trees.[52]

During a 1983 regional forestry meeting, Darrow's colleagues concurred that the eucalyptus plantation projects in Ovamboland had not been very successful. But they attributed the limited success primarily to the late 1970s and early 1980s droughts that plagued the region rather than to the unsuitability of eucalyptus or the poor scientific quality of the trials. The rationale that the poor performance was because of the vagrancies of Nature led to plans to greatly expand the Onankali plantation. Only the end of South African colonial rule and the independence of Namibia in 1990 relegated the plans to a dusty drawer.[53]

Despite having demonstrated that eucalyptus trees could be grown in Ovamboland, the results of the trials had been mixed at best. In fact, during the early 1990s, eucalyptus trees could be seen not only on the plantations (however poorly they looked) but also occasionally as an on-farm tree. In fact, 15 percent of the respondents to the 1993 OMITI survey mentioned having planted eucalyptus seedlings, while 17 percent of the respondents listed eucalyptus among trees that they desired, particularly as shade trees. A local nursery anticipated an annual demand of 800 eucalyptus treelings.[54] Considering the objectives for establishing the forest plantations, however, they were a resounding failure. Ovamboland's

52 K. Darrow, "Provenance-type Trials of Eucalyptus Camaldulensis and Eucalyptus Tereticornis in South Africa and South-West Africa: Eight-year Results," *South African Forestry Journal*, 124 (1983), pp. 13–22. See also R. Hilbert, "Importance of Provenance Research and Tree Breeding to Achieve Higher Yields" (Paper presented at the 10th regular meeting of the SARCUSS Standing Committee for Forestry held at Sabie, RSA, May 20–24, 1985).

53 NAO, OVA 56, Report Foresters Meeting Rundu, March 24, 1983. Kreike, "An Inventory of Trials with Exotic Tree Species in Northern Namibia." The Opono lake trial plot was established in 1968, AHE 1/342, Quarterly Report District Forest Official, December 30, 1968.

54 OMITI 4.4.12, 4.4.30. The nursery was the DAPP nursery at Ombalantu, Sheuyange Tufaneni (second-year student, Agriculture, Ogongo Agricultural College) report of a field trip to the DAPP nursery, Ombalantu, n.d. [July 1993].

population required not the timber trees the foresters sought to grow but poles and sticks. However, not a single farm or hut was constructed with eucalyptus poles because it proved to be extremely vulnerable to termites and other local pests. To serve any function at all, the wood had to be treated with insecticides before and preservatives after it was harvested.

Conclusion

Although colonial water technology and resource management mitigated the impact of environmental constraints and/or enhanced environmental opportunities, in the process affecting the environment, humans did not gain full dominance over Nature. Eastern Ovamboland, e.g., is a mosaic environment with stark contrasts. The hydrology of eastern Ovamboland prevented the development of a dense settlement pattern: in contrast to the floodplain where villages closely abut one another, villages in eastern Ovamboland are often separated by large stretches of forest. The local character of environmental change serves to further complicate linear models of environmental change because it suggests that environmental change does not necessarily "blanket" over whatever previously existed, but instead develops like a patchwork. Moreover, different patches may experience different patterns of environmental change, and at different speeds.

Ovamboland's canal system was state-built, state-controlled, and state-managed. It also was originally designed to tame the floods as much as it aimed to harvest water. Operating within a modernization paradigm, Ovamboland's colonial rulers considered water management based on scientific technology (irrigation) to be the key to halting environmental degradation and to jumpstarting economic development in Ovamboland. Colonial dams also fit within the classic hydraulic society analysis. But the transformation of the water infrastructure that took place in the Ovambo floodplain was neither an accomplishment of Western science nor strictly a product of the colonial era. Instead, colonial science and technology built on and interacted with the local water harvesting system; and, in the 1990s, its core technology – water holes – were still for many households critical year-round sources of water. In brief, the colonial administration and its engineers failed to bring water management in Ovamboland under centralized scientific control. The thousands of water holes that continued to exist were individually owned and managed. In this respect, Ovamboland's character as a decentralized hydraulic society is by no means unique: the elaborate water management systems of other African societies, e.g., that of the Shambaa of Tanzania, as well as those of ancient India and the early modern era Low Countries were managed by local communities (cities, water boards), and not by a centralized state.[55]

The history of water harvesting in Ovamboland in the twentieth century thus addresses the issue of levels of analysis as well as the question of dominance. In environmentally deterministic analysis, challenging hydraulic environmental conditions (e.g., water scarcity or flooding, or both), and the resultant need for intensive water management sometimes is seen to give rise to elaborate and highly centralized societies. Moreover, despite their environmentally deterministic origins, the hydraulic societies that are the product of the specific human–nature interaction are considered to be emblematic of the triumph of human dominance, or Culture over Nature, outclassed in this respect only by urban societies. But, the main sources of water were not constructed, controlled, or managed by the state.

While colonial hydrology experts from the 1930s through the 1960s eagerly expanded on and made use of local technology and knowledge by locating large water reservoirs on the sites of water hole systems and by sponsoring well digging, colonial forestry experts in the 1960s through the 1980s took the opposite tack. Scientific forestry in Ovamboland aimed to create large state-controlled plantations with fast-growing, seed-grown exotics. The eucalyptus plantations, however, were hardly examples of good science. Experiments were executed inconsistently and the foresters involved failed to maintain proper documentation: subsequent follow-up assessments of the trials to measure tree growth and the survival rate relied on a substantial dose of guesswork because researchers had no records about which trees had been planted and where, or what inputs were used. In addition, despite the rhetoric that they were to provide local fuel and construction wood, the plantations in fact were conceived and managed to produce conventional timber trees. Ovambo farms and palisades, however, did not require the timber-size trunks produced by full-grown trees but poles and sticks from trees and bush. Although they successfully reforested in terms of output, in terms of use, the eucalyptus

55 On Shambaa irrigation, see Feierman, *Peasant Intellectuals*, pp. 64–65. On ancient India, see R. Chakravarti, "The Creation and Expansion of Settlements and Management of Hydraulic Resources in Ancient India," in Grove, Damodaran, and Sangwan, *Nature and the Orient*, pp. 87–105. On the Low Countries, see Van der Ven, *Leefbaar Laagland*, pp. 33–152.

plantations were a dismal failure: not a single homestead or palisade was constructed using plantation-grown wood.

Overall, post–World War II colonial development projects caused important environmental changes, especially in how they affected settlement in Ovamboland by facilitating urbanization and in terms of how they affect the region's hydrology, e.g., by dramatically increasing dry season water availability along the canals and pipelines and decreasing flood water availability south of the canals. Colonial engineers failed to conquer Nature but did participate in processes of environing, highlighting that the concept may be equally useful to "modern" and/or "Western" societies as to "pre-modern" and/or "non-Western" societies as a tool to move beyond the long shadow of the Nature–Culture dichotomy.

Conclusion

Dense settlement in Ovamboland's semiarid climate was dependent on an extensive environmental infrastructure that humans crafted and shaped in conjunction with prevailing environmental resources and forces. The region's towering fruit trees were neither wild nor wilderness trees. To the contrary, the marula, palm, and birdplum trees marked village environments. The palisades and fences that farmers erected around their homesteads after they cleared the preexisting woodlands protected the fruit trees from animal predation. Farmers mounded and ridged the soils below the fruit trees, thereby redistributing on-farm soil resources. In addition, farmers and their animals imported organic matter and other nutrients from beyond the fields and consequently influenced the physical and chemical processes of soil formation. On-farm soil enrichment drained nutrients from the surrounding off-farm soils, impoverishing them. Ovamboland's remaining woodlands – located on the edges and between the villages – were able to meet a high demand for fire and construction wood mainly because of mopane bush coppice, which sharply reduced woody species diversity. The thousands of water holes and reservoirs that explorers, missionaries, and officials described were not *natural* resources because they did not occur naturally. Instead, households harvested water from local catchment areas and stored the water on-farm for use during the dry season. By the early 1990s, the majority of Ovamboland's households were dependent on human-dug water holes. But the water holes did not represent human domination over Nature, for they drew on numerous highly localized and low-yielding subterraneous water lenses. To preserve water supplies for human and smallstock use only, households relocated their precious cattle to remote cattleposts for the duration of the dry season. Beyond the villages, browse and burn were the chief tools to create and maintain numerous patches of bush savanna pasture connected by long transhumance trails. These disparate tools and practices resulted in an elaborate environmental infrastructure. The concept of environmental infrastructure and its relation to processes of environmental change thus provides the framework for a reconceptualization of environmental change across time and space. Drawing on the history of twentieth-century north-central Namibia, the study argues that the dynamics of the production and the reproduction of environmental infrastructure by human (and nonhuman) architects of Nature advance understanding of environmental change that moves beyond the linearity, homogeneity, and hierarchy of existing models.

Processes of *Environing*

Ovamboland's environmental infrastructure was never merely a product of a one-time investment of labor, capital, and other resources that changed the nature of the region's environment forever. Environmental infrastructure is not a static outcome. To the contrary, continuous processes of environing encapsulate multiple histories of environmental change. Fruit trees had to be protected against animals. Any failure to maintain fields and their soils – by clearing, burning, cultivating, weeding, manuring, cropping, pasturing, and fallowing – changed not only soil composition, but also the vegetation cover. Coppice woodland required repeated management and use; cutting bush too short or too often or ceasing coppice management altogether changed its form, product, and nature. Water holes and reservoirs had to be re-dug and cleaned on a regular basis, requiring investments not only by their owners but also by wider networks of users that ranged from several adjacent households (in the case of water holes) to neighborhoods and even entire villages (in the case of reservoirs). In the absence of burn and browse, dense bush quickly invaded farms, fields, and savanna.

Ovamboland's towering trees, black soils, water sources, and coppice and browse bushlands thus were not simply managed *natural* resources and evidence of a harmonious or sustainable Nature–Culture relationship. They also were not mere cultural artifacts; however, because they were neither inanimate tools to dominate Nature nor the product of human domination over Nature. Ovamboland's inhabitants worked *with* Nature (not *within* or *outside* Nature), shaping the local environment, e.g., by constructing water holes to tap local water lenses. Working with

Nature does not imply that environmental infrastructure was or is by definition in harmony with Nature or sustainable. Processes of environing in twentieth-century Ovamboland continuously changed the environment, with consequences both negative and positive: trees were felled, burned, coppiced, protected, and planted; on-farm soils were enriched while off-farm soils were impoverished; and water was drained, diverted, polluted, harvested, consumed, and conserved.

Ovamboland's inhabitants worked *with* Nature, interacting with animals, local rainfall, regional run-off, and other resources, while creating, maintaining, and re-imagining environmental infrastructure. The resulting environmental infrastructure was highly dynamic, not only because of the contributions of a myriad of agents and forces, but also because the processes of environing were as continuous as they were variable. Climate and weather patterns, e.g., affected environmental infrastructure directly, resulting in droughts and floods that required repairs and adjustments to meet prevailing challenges, both old and new. Changes in climate and weather patterns also impacted environmental infrastructure indirectly, by shaping peoples' actions, needs, and perceptions. In addition, political, social, economic, and a multitude of other factors influenced how people physically and mentally mapped, shaped, and reworked environmental infrastructure. Environmental infrastructure is thus simultaneously outcome and process, because it is always *in process*.

Emphasizing environmental change as a process diminishes the danger of its application as a static monolith. Environmental infrastructure is neither a state, nor a condition of more or less environmental infrastructure, or more or less environing. Instead, the concept of environing explicitly acknowledges that humans, other organisms, and inorganic matter always influence their environments, even as they are in turn part and parcel of that same environment and are shaped by it themselves. In addition, the co-shaping process is constant and involves multiple moving targets, if only because it involves multiple agents making decisions based on motivations and urges that in turn are subject to change over time and in space. Co-agents, whether organic or inorganic, have their own "objectives" and dynamics. Multiple processes of environmental change may unfold simultaneously or in dissynchronous fashion, and they may interact through feedback mechanisms or remain compartmentalized to varying degrees. Environmental infrastructure and environing first and foremost offer analytical and descriptive tools – particular lenses – to move beyond the Nature/Culture, non-West/West, and pre-Modern/Modern dichotomies. In that sense, the concepts appear ahistorical: they can be applied through out history. They are also deeply historical concepts, however, because they emphasize process over outcome and multiple dissynchronous subprocesses of environmental change, even within a defined time and place. These processes are ongoing: the present is merely a fleeting moment.

Individual households managed environing, initiating different environmental processes at different times. For example, the first household to settle in a new village site would have a head start on subsequent arrivals in terms of clearing the land, digging water holes, constructing huts, palisades, and fences, building soils, coppicing the bush, and propagating fruit trees. Processes of environing therefore often were out of sync, even within a single village. For example, three or four decades into the founding of a new village, the first settlers' farms may have been marked by towering fruit trees, while the plots of recent arrivals were mostly denuded and barren. The multiple processes of environing also interacted: the environmental infrastructure that the first settlers in each individual village had created often critically sustained subsequent settlers, who relied on the water holes that the pioneers had dug. In eastern Ovamboland, the founders of new villages often enjoyed the use of a preexisting infrastructure of water holes constructed by Ovambo and San hunters and herdsmen.[1] As the herdsmen and hunters moved further east, the farmers followed, grafting onto ongoing processes of environing and gradually appropriating the water holes. The colonial alienation of water hole complexes to construct water reservoirs ("dams") interrupted these previously ongoing processes of environing, even as it added yet another layer of environing that served to jump-start colonial projects.

The intent of the concept of environing is twofold. First, it explicitly excises the Nature–Culture dichotomy and its analytical and narrative conceptualization and vocabulary from the equation as the standard and scale with which to measure environmental change. Although the Nature–Culture dichotomy has been soundly rejected by leading environmental historians since the late 1990s, it remains deeply embedded in how environmental change is analyzed and presented because of the continued use of language and vocabulary that conceptually is defined from and by the dichotomy, including, e.g., the domestication-related vocabulary. In this respect, one concern may be that excising

[1] See Kreike, *Deforestation and Reforestation*, pp. 168–176. Scarborough stresses that water holes were critical for hunter-gatherers from South Africa to Australia. See Scarborough, *The Flow of Power*, p. 9.

the Nature–Culture dichotomy robs environmental and environmentalist narratives of many of their dramatic and rhetorical powers. The study of environmental change is not only an academic exercise. In addition to being history, environmental history and related fields that emphasize the human–environment interaction also aim to *remake* history. Environmental history has an emancipatory and liberating agenda in common with, e.g., Africanist history, African-American studies, subaltern studies, peasant studies, and studies of resistance, indigenous peoples, gender, and sexuality (i.e., ending the destruction of the environment on which we all depend). Without a clear pre-contact, or precapitalist "pristine" Nature baseline with which to measure environmental decline, and without the power of a Nature-to-Culture "crisis" narrative, it becomes more challenging to push an action agenda for saving the environment, e.g., by holding up past environmental collapes or successes. The quest for an action agenda explains in part why the Nature–Culture dichotomy persists: it offers a clear narrative arc for a good and noble case that requires urgent remedy. Even if the cause justifies the means, however, it should not prevent a critical reevaluation of the weaknesses of the approach.

Second, and perhaps more controversial, the concept of environing problematizes the notion that there is an a priori and fundamental structural and qualitative difference in how premodern and modern, Western and non-Western, and human and nonhuman agents affect and in turn are affected by environmental change. New Orleans city-dwellers in the early twenty-first-century Mississippi delta proved no less immune to floods than their contemporaries in Ovamboland and Bangladesh, sixteenth-century Dutch farmers in the Rhine Delta, or twelfth-century Chinese farmers in the Pearl River Delta. But the populations of non-Western or premodern Ovamboland, Bangladesh, the Netherlands, and China were protected at best by earthen mounds and dikes while New Orleans' population was shielded by concrete levees designed and constructed by the U.S. Army Corps of Engineers using the latest science and technology. Moreover, dependent on a dwindling and nonrenewable hydrocarbon resource, early twenty-first-century North Americans are as vulnerable to oil price hikes as are rural Angolans who rely on bush taxis and kerosene lighting.

The commodification of Nature is often identified as a uniquely "Western" contribution to the mechanics of environmental change, and it is linked to the rise of global capitalism. In Asia, however, the commodification of Nature predated Western capitalist and imperial intrusion. In similar vein, Isenberg argued that the transformation of the western plains Amerindians from hunting and gathering generalists to horse-mounted, bison-hunting specialists who produced hides for the market was not a linear and mechanistic reaction to the expansion of Western capitalism alone, but the outcome of complicated local and extra-local social and economic processes.[2] That capitalism is not the only explanation behind the unbridled and destructive exploitation of the twentieth-century global environment is borne out by the equally abysmal environmental record of state socialism.[3] Significantly, McNeill suggests that modern Western industrial society may not be fundamentally different in its qualitative environmental impact from precapitalist societies, pointing to the previously mentioned case of imperial China. McNeill contends that it was the *scale* of the environmental change that occurred – because of the dramatic population growth, mass consumption patterns, and massive application of technology – that made the twentieth century an era without precedent in human history.[4] Analytically distinguishing the processes of environmental change in Western/modern societies as qualitatively different from those in non-Western/premodern contexts may have some merits, but it is also artificial, arbitrary, and sometimes entirely misleading. Reimagining environmental change in both Western/modern and non-Western/pre-modern contexts in terms of environing provides an alternative lens beyond the rehearsed dichotomies.

The concept of environing, e.g., tones down the differences and the transitions between what often are still represented as evolutionary stages of human development as distinguished by the degree of control a society wields over Nature: hunting and gathering (living by Nature), food production (agriculture and pastoralism through the domestication of plants and animals [i.e., harnessing or taming Nature]), and industrial society (artifice and domination over Nature). The West–non-West, modern and premodern, and human–animal dichotomies are intimately connected to the idea of distinct and progressive evolutionary stages.

[2] Marks, *Tigers, Rice, and Silt*, pp. 248–276 and 338; G. Bankoff and P. Boomgaard (eds.), *A History of Natural Resources in Asia*, pp. 1–18; J. W. Christie, "The Agricultural Economies of Early Java and Bali," pp. 47–67; and Isenberg, *The Destruction of the Bison*.

[3] See, e.g., Scott, who stresses that high modernist projects seeking to dominate Nature were the hallmark of both capitalism and communism; see Scott, *Seeing Like a State*.

[4] McNeill, *Something New Under the Sun*, and Marks, *Tigers, Rice, and Silt*, pp. 248–276 and 338.

Non-West and premodern were considered to be more primitive and closer to the state of Nature; similarly, primitive humans (as opposed to civilized humans) and animals (as opposed to humans) lived by instinct (and in and by Nature) rather than by reason (and in and by Culture).[5]

The transitions between evolutionary stages were depicted as revolutions.[6] Proponents of modernization and its malcontents regarded hunter-gatherers as living in and by Nature, while industrial society stood for the domination of Culture over Nature. Environmental malcontents – declinists and inclinists – accepted the Nature-to-Culture linear narrative. They inverted, however, the grading scale imposed by modernizers: the closer a society's modes of livelihood were to the state of Nature/Primitiveness (as with hunter-gatherers), the less environmentally destructive a society's ideas and practices. Whereas in the modernizing perspective, each successive environmental revolution improved the condition of humanity, its critics argued that each revolution, in fact, wrought more destruction on ecosystem earth, with long-term negative consequences for human welfare and survival. This was and is a very important debate: obviously human society cannot be sustained by adopting a hunter-gather lifestyle, but it is equally evident that we cannot continue to follow the current path. However existential and important the debate, conceptually and narratively it keeps us running in the same Nature–Culture rut.

Environing functions as a tool to "unstuck" the analysis. Hunting and gathering, pastoralism, agriculture, and industrialism all can be seen in terms of environing: hunters and gatherers, prehistorically, historically, and currently shape their environments, and industrial and postindustrial societies continue to be dependent on and vulnerable to floods and droughts, epidemics, and global climate change. Rather than using what essentially are modes of production as successive evolutionary stages, environmental infrastructure allows for combinations of different livelihood forms to occur at the same time without expressing a priori value judgment about the society's evolutionary and social development and its

[5] On the human–animal boundary, see, e.g., E. L. Milam, *Looking for a Few Good Males: Female Choice in Evolutionary Biology* (Baltimore, MD: Johns Hopkins University, 2010).

[6] Cohen and Crane-Kamer emphasize that changes in food production were much more gradual and did not always constitute an unambiguous improvement: the transition to maize cultivation in the Americas, e.g., has been associated with an increase in health risks. See M. N. Cohen and C. M. M. Crane-Kramer (eds.), *Ancient Health*.

attendant environmental impact in terms of destruction, improvement, or sustainability.[7]

Although this study focuses on humans' role in environing, it also highlights the contributions of nonhuman agents. Fauna and flora play as critical a role in processes of vegetation and soil formation as humans. Cattle, goats, termites, and beavers engage in environing. So do microbes; witness the histories of plague, smallpox, malaria, and cholera. The same is true for plants, especially "invasive" species that colonize lawns, pastures, islands, and entire continents despite the massive use of herbicides.[8] Inanimate agency adds yet another dimension: the roles of geological forces, gravity, climate, weather, water, and fire can hardly be denied. Thus it is not surprising, e.g., that farmers in both Namibia and Nigeria perceive that soils are by turn lean and fertile, that soils wax and grow, change shape and color, and that they move (through erosion).[9]

Operationalizing *Environing*

Operationalizing the concept of environing requires raising the questions of what is changing, for and by whom or what, and why. It also requires an emphasis on the process of change, rather than a fixation on

[7] Scott rejects the evolutionary hierarchy of what he identifies as different contemporary strategies to escape or impose the state. He describes hunting and gathering and shifting cultivation, e.g., as state-evading strategies, because mobile societies are difficult to control and tax by the state. See Scott, *Seeing Like a State* and Scott, *The Art of Not Being Governed*.

[8] Ritvo points out that in Western Europe, animals were seen as actors well into the early modern era. She demonstrates that it was only in the nineteenth century that animals lost their status as agents and were no longer held criminally liable for actions they were involved in. See Ritvo, *The Animal Estate*, pp. 1–42. On beavers, see Vileisis, *Discovering the Unknown Landscape*, p. 16. Although Darwin accepted animals as agents (especially in terms of sexual choice), the idea that animals had agency was overwhelmingly denied during the latter part of the nineteenth century and most of the twentieth century. See Milam, *Looking for a Few Good Males* and Ritvo, *The Animal Estate*. The concept of "ecosystem engineers" emphasizes that organisms create "artifact." See C. G. Jones, J. H. Lawton, and M. Shachak, "Organisms as Ecosystem Engineers," *Oikos*, vol. 69 (April 1994), no. 3, pp. 373–386. See also K. Berkenbush and A. A. Powder, "Ecosystem Engineering: Moving Away from 'Just so' Stories," *New Zealand Journal of Ecology*, 27 (2003), pp. 67–73.

[9] For soils as "actors," see McCann, *Green Land, Brown Land, Black Land*, p. 12. Ishida, Tian, and Wakatsuki, "Indigenous Knowledge and Soil Management," pp. 91–109 (especially 196–197). McNeil and Winiwarter stress that soils have both their own natural and human histories, see J. R. McNeill and V. Winiwarter, "Soils. Soil Knowledge and Environmental History: An Introduction," in J. R. McNeill and V. Winiwarter (eds.), *Soils and Societies: Perspectives from Environmental History* (Isle of Harris, UK: The White Horse Press, 2006), pp. 1–6.

outcome(s) and baseline(s), and the acknowledgment that the processes and the objects and subjects involved in the changes are differentiated in composition, time, and place. The various processes may intersect and interconnect, but they also are subject to their own causative, spatial, and temporal dynamics. All people, animals, and other subjects and objects in processes of environmental changes are not equally affected and affecting: in the case of humans, e.g., class, race, ethnicity, gender, and other ascribed characteristics shape how individuals and groups are affected by the "environment" and how they in turn shape their environment.[10] Moreover, environmental change and environing cannot be understood in terms of a singular process; instead, it involves multiple (sub)processes. Identifying the processes and sub-processes that occur, the extent to which they interact or not and why, and who and what affects whom, what and why, as well as who and what benefits or loses and why, highlights the dynamics of environmental change beyond a singular and unilinear trajectory. Conventionally, human observers, actors, or subjects perceive a singular historical process of environmental change differently, based on their consciousness, experiences, or relative position; that is, multiple mental interpretations evolve from what is perceived to be a single historical process. The argument here, however, is that numerous historical (sub)processes may be co-occurring, simultaneously or out of sync with one another.[11] The various processes may also be compartmentalized from one another and/or interactive, allowing for environmental changes to occur at different speeds. The potential for rapid, even "revolutionary" change exists, especially when different processes intersect and interact.[12]

In addition, the central and critical research question in the study of environmental change should *not* be to determine the extent to which an

[10] Differentiating the subjects and objects of environmental change beyond the categories West–non-West, and (colonial) rulers-subjects is an important contribution offered by the environmental justice literature to the study of environmental history. A classic is A. Hurley, *Environmental Inequalities: Class, Race, and Industrial Pollution in Gary, Indiana, 1945–1980* (Chapel Hill: University of Chapel Hill Press, 1995). For a rare and path-breaking explicit application of the concept – developed in the context of American environmental history – to a non-Western case, see Jacobs, *Environment, Power, and Injustice*.

[11] I have also made the argument earlier in Kreike, *Deforestation and Reforestation*.

[12] An emphasis on "process" may be construed as highlighting gradual change over rapid change. Criticizing the linearity and "gradualism" inherent in perceptions of change in the fields of ecology and evolutionary biology, Blumer argues that the origins of agriculture may not have originated in the form of a gradual process. Instead, he argues, the transformation to agriculture may have occurred in short bursts driven by specific events, see M. A. Blumer, "Ecology, Evolutionary Theory and Agricultural Origins," in

environment has been degraded, improved, or sustainably used, because this approach foregrounds *judging* the process a priori, by focusing on the outcome. Instead, the initial questions should be: what is changing; who or what drives the change(s); and what are the effects? Only once the processes of change have been analyzed and described can the impact of the changes be evaluated.[13]

In a post–Nature–Culture dichotomy era, environmental change can no longer be construed in terms of a linear Nature-to-Culture trajectory. Moreover, environmental change can no longer be explained in terms of two discrete categories of forces subject to their own distinct dynamics (i.e., with Nature being subject to natural laws and Culture subject to the actions and imaginations of humans). Environing offers an approach that conceptually and methodologically de-centers the continuing legacy of the Nature–Culture dichotomy, moving beyond clear-cut distinctions between the object and subject of environmental change and facilitating an analysis of the multiple synchronous and dissynchronous processes that also may occur. The result is a richer and multidimensional description and analysis of environmental change.

The concept of environing is not merely utilitarian in the sense that Nature or "natural resources" only are valuable through association with humans. It is also a reflection of a reality that has its origins in the birth of humankind: human society and environment, Nature and Culture, are, were, and will remain intricately and intimately intertwined. In the late twentieth century, it seemed that humans had conquered Nature or caused its death: science and technology had made the natural world irrelevant for the future and its preservation seemed to be only of aesthetic, moral, or antiquarian concern. In the early twenty-first century, however, this perspective has changed radically as a result of, for example, the challenges of global climate change and the opportunities offered by biotechnology.

Western and non-Western and modern and traditional farmers, peasants, laborers, rulers, soldiers, and engineers do not manage and use

D. R. Harris (ed.), *The Origins and Spread of Agriculture and Pastoralism in Eurasia* (Washington, DC: Smithsonian Institute Press, 1996), pp. 25–49.

13 Radkau warns against rushing in with value judgments and privileging the outcome; see Radkau, *Nature and Power*, pp. 20–22, 26–27, and 35. McNeill, Blackbourn, Jacobs, and Radkau all call for a more empirical focus on environmental change, rather than working from a hypothesis of environmental degradation or improvement. See Jacobs, *Environment, Power, and Injustice*, p. 218; Radkau, *Nature and Power*, pp. 20–22; Blackbourn, *The Conquest of Nature*, pp. 9–13; McNeill, *Something New Under the Sun*.

natural resources or create *human* artifact, but physically and mentally re-create, maintain, change, or destroy *environmental infrastructure*. Environmental infrastructure encompasses both Nature and Culture and human (and nonhuman) environmental actors constitute architects of Nature working with, through, and alongside Nature, even as they seek to dominate Nature. Nature is neither pristine nor dead; environmental infrastructure is not by definition *green* infrastructure (as in environmentally friendly, sustainable, or in harmony with Nature); and environing often involves abusing and polluting the environment, which is all the more dangerous because we do not fully understand the processes involved. Removing, adding, or changing environmental forces or actors (e.g., through plant or animal extinction or their confinement to limited spaces) may critically reorient the processes of environmental change and transform the environmental infrastructure on which other actors depend; e.g., rising sea levels as a result of global climate change constitute a serious challenge to both "traditional" and "modern" coastal water management infrastructure worldwide.

The plurality of environing as a concept has implications for power and ownership relationships – that is, it shapes the political ecology. In a Lockean context, the improvement of a natural resource through domesticating, harnessing, and enhancing its utilitarian value, creates exclusive property rights. This notion is a core tenet of the conquest-of-Nature narratives of modernization and development, capitalist *and* non-capitalist, including, but not confined to, its colonial and imperial manifestations.[14] The concepts of Nature (as pristine wilderness and the "other" of Culture) and domestication (as a linear Nature-to-Culture progression) are highly problematic and the Lockean-derived notion of property ultimately is embedded in a Nature–Culture dichotomy. The notion is based on the premise that change occurs in the form of a singular process, and it privileges outcome over process. It also presumes that there is clear agency. With multiple agents engaging in multiple processes, and the blurring of boundaries between agents and tools, and subjects and objects, however, questions of who or what owns or rules over whom or what become very difficult to answer. If Ovambo villagers grafted their environmental infrastructure on the preexisting environmental infrastructure of San and Ovambo hunters and pastoralists, whose environmental infrastructure is it? Humans have used domestic animals, fire, and water as tools for thousands of years. Are cattle and goats, cultivated plants, and fire and water mere human tools? How interdependent are animals, plants, fire, water,

[14] See, e.g., Graham, *Lawscape*, pp. 84–133.

and humans? Did humans domesticate livestock, crops, and diseases, or did livestock, crops, and diseases domesticate us?[15] The environing concept allows for the pluralities, ambiguities, intricacies, and multidimensionalities of environmental processes.

The preceding discussion emphasizes how environmental change serves humans, who simultaneously have been highlighted as major environmental *infrastructurers*, and, in that sense, the concept appears anthropocentric and anthropogenic.[16] The concept of environmental infrastructure, however, acknowledges both human and nonhuman agency and moreover conveys that humans as well as nonhumans are simultaneously critically dependent on that infrastructure. Ironically, the term infrastructure is somewhat ambiguous; it implies both a measure of human control (because it is an artifact), *and* human dependency, because it provides the basis for, enables, and facilitates human existence.

The study also stresses that humans are not the sole creators, users, and managers of environmental infrastructure. Nonhuman actors, forces, and dynamics are critical co-creators, co-users, and co-managers, thence the rejection of the concepts of Culture (as opposed to and as in domination over Nature) and domestication. To argue that there is neither any real Nature left to save nor any true Culture to preserve does not mean that humanity should continue on its present path. To the contrary, the awareness that humanity is far from gaining a full understanding

[15] Totman refers to agriculture in Japan in general and the cultivation of the *sugi* tree (*Cryptomeria japonica*) in particular as "symbiotic alliances with collaborating species of plants and animals," causing radical ecological change and enabling the *sugi* trees and other species to establish hegemony. Totman sees this as resulting in a new ecological equilibrium. The ideas of symbiosis and equilibrium are too suggestive of a harmony model within a Nature–Culture dichotomy, but the idea that humans and other species use one another – however uneven their relationship might be – is very valuable. See Totman, *The Green Archipelago*, pp. 186–187.

[16] Although ancient philosophers, notably Aristotle, are often referred to as the source of the idea that Nature's purpose is to serve humanity, Hughes notes that there was no consensus on this issue in Antiquity. In fact, Aristotle's own student Theophrastus disagreed with his teacher that everything in Nature was arranged hierarchically to serve humans' needs. Instead, Theophrastus argued that it was not always clear who served what and posited that nonhuman creatures existed not merely to serve humanity. The followers of Pythagoras took a more radical position and forbade the killing of animals or plants for human use. See Hughes, *Pan's Travail*, pp. 55–62. Modernization and inclinist narratives tend to be anthropocentric while declinist narratives are often ecosystem-centric. Tiffen, e.g., distinguishes two schools of environmental degradation: one concerned with the natural environment and species diversity and another that focuses on maintaining people's living standards as the main yardstick for environmental sustainability. See Tiffen, *More People, Less Erosion*, p. 14.

of – let alone full control over – the environment on which we depend for survival should make us tread even more lightly. Moreover, because neither individual humans nor humanity as a whole dominates the multiple processes of environing, and because the myriad existing environmental infrastructures serve a plethora of human and nonhuman constituencies, these infrastructures by definition are *shared* resources. As humans, we are bound not only to exercise our own diverse environmental rights and obligations but also to consider the interdependent role of organic and inorganic entities in shaping our shared environment.

Is everything environmental infrastructure? In a sense, it is because everything ultimately is the (co-)creation of human and nonhuman actors, forces, and processes. Everything is also environmental infrastructure because everything serves to support or facilitate processes that shape animate and inanimate objects or substances, even factors and conditions that we see (rightly so) as degrading our environment, including, e.g., wild fires, floods, bush encroachment, deforestation, air and water pollution, and global warming. Wild fires and floods are destructive for many species and biological communities, even if they also create favorable environmental conditions for, respectively, fire- and water-tolerant species and communities. Deforestation not only facilitates human settlement and agriculture but also encourages grasses and herbs, creating the conditions to sustain large herds of livestock and wildlife. Thermal water pollution favors algae growth, and rising CO_2 levels in the atmosphere promote plant growth (through carbon sequestration). Environmental infrastructure and environing are neither by definition nor a priori good or bad: instead, the concepts shift the emphasis of the analysis, description, and evaluation of environmental change from a fixation with outcome and baseline to a more balanced approach that focuses on the dynamic and intricate processes that shape our world.

Index